# DISASTER ON THE HORIZON

## HIGH STAKES, HIGH RISKS, AND THE STORY BEHIND THE DEEPWATER WELL BLOWOUT

## BOB CAVNAR

Chelsea Green Publishing
White River Junction, Vermont

Project Manager: Patricia Stone
Editorial Contact: Joni Praded
Developmental Editor: Joni Praded
Copy Editor: Laura Jorstad
Proofreader: Nancy Ringer
Indexer: Peggy Holloway
Designer: Peter Holm,
    Sterling Hill Productions

Printed in the United States of America
First printing October, 2010
10 9 8 7 6 5 4 3 2 1     10 11 12 13 14

## Our Commitment to Green Publishing

Chelsea Green sees publishing as a tool for cultural change and ecological stewardship.
We strive to align our book manufacturing practices with our editorial mission and to
reduce the impact of our business enterprise in the environment. We print our books and
catalogs on chlorine-free recycled paper, using vegetable-based inks whenever possible.
This book may cost slightly more because we use recycled paper, and we hope you'll agree
that it's worth it. Chelsea Green is a member of the Green Press Initiative (www.green
pressinitiative.org), a nonprofit coalition of publishers, manufacturers, and authors working to protect the world's endangered forests and conserve natural resources. *Disaster on
the Horizon* was printed on Rolland Enviro Natural, a 100-percent postconsumer recycled
paper supplied by Thomson-Shore.

**Library of Congress Cataloging-in-Publication Data**
on file with the publisher

Chelsea Green Publishing Company
Post Office Box 428
White River Junction, VT 05001
(802) 295-6300
www.chelseagreen.com

# In Memoriam

As I was writing this book, recalling my own experiences during my career in the oil field and recounting the events surrounding the disaster on the *Deepwater Horizon*, I was constantly aware of those who were lost on that tragic night of April 20, 2010. We should always keep them and their families in our thoughts and prayers.

**Jason Anderson,** 35, toolpusher, of Midfield, Texas. He left behind his wife, Shelley, and two children.

**Dale Burkeen,** 37, crane operator, from outside Philadelphia, Mississippi. He left behind his wife, Rhonda, and two children.

**Donald Clark,** 49, assistant driller, of Newellton, Louisiana. He left behind his wife, Sheila, and four children.

**Stephen Curtis,** 40, assistant driller, of Georgetown, Louisiana. He left behind his wife, Nancy, and two children.

**Gordon Jones,** 28, mud engineer, of Baton Rouge, Louisiana. He left behind his pregnant wife, Michelle, and a son. Michelle gave birth to a son 24 days after the blowout.

**Roy Wyatt Kemp,** 27, derrick man recently promoted to assistant driller, of Jonesville, Louisiana. He left behind his wife, Courtney, and two daughters.

**Karl Kleppinger,** 38, floor hand, of Natchez, Mississippi. He left behind his wife, Tracy, and a son.

**Blair Manuel,** 56, mud engineer, of St. Amant, Louisiana. He left behind his fiancée, Melinda, and three daughters.

**Dewey Revette,** 48, driller, of State Line, Mississippi. He left behind his wife, Sherri, and two daughters.

**Shane Roshto,** 22, floor hand, of Franklin County, Mississippi. He left behind his wife, Natalie, and one child.

**Adam Weise,** 24, floor hand, of Yorktown, Texas.

*When he shall die*
*Take him and cut him out in little stars*
*And he will make the face of heav'n so fine*
*That all the world will be in love with night*
*And pay no worship to the garish sun.*
—William Shakespeare, *Romeo and Juliet*

José Arcadio Buendia dreamed that night that right there a noisy city with houses having mirror walls rose up. He asked what city it was and they answered him with a name that he had never heard, that had no meaning at all, but that had a supernatural echo in his dream: Macondo.

—GABRIEL GARCÍA MÁRQUEZ,
*One Hundred Years of Solitude*

# CONTENTS

*Phoom!* It was that impossible-to-describe sound that happens when you're too close to the blast to hear the full roar.

I could see nothing but orange, black, and red as I was engulfed in an intense, all-consuming heat. It was immediately sickening, and I began to process the fact that I was in serious trouble. At that same instant, I was knocked off my feet and slid headfirst down a muddy slope, plunging into water saturated with chemicals. The fire was overhead now, and I was completely disoriented. Crawling on my belly, underwater, unable to breathe, I was trying to get away from the heat that had permeated my skin. I crawled for what seemed like hours, though it was only seconds—everything in slow motion.

Suddenly I felt strong hands grab me, pulling me from the ditch I had been blown into. I was doused with ice-cold water from a drinking-water can, causing me to draw a sharp breath. I was alive. Ken, an oil field truck driver I knew well, was shouting in my face, asking if I was okay. But I couldn't see him clearly or hear a thing. As my breathing improved, I came around and became aware that I was almost completely undressed. The blast had blown off most of my clothes, including my boots, hard hat, watch, and the sunglasses that had protected my eyes. My jeans were around my ankles. I was in one piece, but definitely a "crispy critter." My mustache and the hair on the front and left side of my head were fried. I was covered in frac gel, and the skin on my face and left side felt like it was on fire. I was lucky that the breath had been knocked out of me, saving my lungs from ingesting flame.

I breathed deeply as Ken doused me with ice water again and smeared me with Silvadene, a commonly used burn cream. He had himself been burned in a refinery fire a few years previously, and he carried a burn kit in his truck. I was lucky. Today I carry little physical evidence from that flash fire. Landing in the ditch had saved me from critical injury. But I think about the incident every day: This brief brush with death changed me forever.

It was 1981. I was 28 years old, superintendent for operations with a small, Dallas-based independent oil and gas producer, and had just survived an up-close-and-personal encounter with a pit fire on a location where I had been completing a Cotton Valley gas well in East Texas. This well, over 11,000 feet deep, had been nothing but trouble from the start. We had beaten our brains out during the drilling phase, stuck pipe multiple times, dealt with drill pipe leaks, and struggled to get production casing to bottom after finally getting the well drilled. On top of hole trouble and mechanical failures, we had to deal with a grouchy landowner who was more than happy to show off the hole he'd blown through his dining room wall with a shotgun, trying to kill his son-in-law. The son-in-law had physically abused the landowner's daughter, and Dad didn't like it. Welcome to East Texas.

Once we got the production pipe (through which gas flows to the surface) set and cemented, we finally got a break. The completion had gone relatively well, at least up until the point that I set myself on fire. The day had started early; I had opened the well right after sunup to flow back the frac job that we had pumped in the previous day. Frac jobs are used to increase the flow of hydrocarbons from a well by pumping into it frac gel, a chemical-laden fluid that carries a coarse proppant to hold open those fractures created by pumping into the productive formation. In those days, air and water pollution were given little thought. It was common practice to use unlined open pits to drain frac and mud tanks and to catch wellbore fluids as the well was drilled and completed. These pits were simply pumped out and then covered up with dirt when well operations were concluded, leaving behind tons of contaminants to leach into the soil. Drilling mud was even spread on pastures in the region, as most ranchers believed that it actually improved the local soil. On this particular day, the well had strong pressure as I opened the frac tree valves and began flowing the fluid directly into the flare pit where frac fluid was recovered. The well began slugging fluid mixed with natural gas as it unloaded. I watched as the well cleaned up and pressure increased, a good sign.

Early in the afternoon, I decided to see if the gas buster, a device installed at the end of the flow line to separate gas from water, would light off. To light the pit flare, I used the technique common at the time: Soak a cotton rag in diesel, light it with a cigarette lighter, stand on the pit edge, and toss it onto the flow line. What I had failed to realize was that the well had produced much more gas and condensate (a very light oil that's basically raw gasoline) than was obvious due to the high volume of water. For several hours, the gas had settled in the pit, being heavier than air, with the condensate coating the water, something I couldn't see, since condensate is clear. The burning rag hit the gas buster . . .

*Phoom!* With me standing right next to it.

I started my career in oil and gas in the mid-1970s, toward the end of the old days of the wildcatters. Most were gone, but a few were still around. The industry, at least domestically, had generally been decentralized during the previous 70 years, made up of thousands of producers that ranged in size from mom-and-pops all the way to the majors like Shell and Exxon. Drilling and completion technology were developing rapidly, but other important elements of the business lagged. Safety programs in many companies were viewed as mere inconvenience and given passing attention, especially by the smaller firms. Since I had begun my own career about six years previous to the fire, I had experienced my own collection of injuries—I'd smashed digits, suffered a broken tooth from being hit in the mouth by a chain on a drilling rig floor, and almost been killed by a falling joint of drill pipe in a near-miss accident on my *first* job on my *first* day on an East Texas drilling rig. That incident was a little too close for comfort.

Missing digits, scars from burns, lacerations, and broken bones were common on every drilling rig floor I had ever stood upon. Drugs and alcohol were the painkillers of choice.

Very early on in my career, I learned that the industry I had chosen, though I loved it, was dominated by the macho myth of big iron, big rigs, wild wells, and wild men. I was swept up in it myself, pushing my own personal limits; my efforts propelled me quickly up the ranks, but my aggressiveness was one of the factors that led up to my losing battle with the pit fire. Rules were made to be broken, and money was A-1. Cash profit was everything, and efforts to make that profit not only pushed the edge of the envelope of responsibility and honesty, but often tore the envelope all to pieces. It was common practice for well-servicing companies to overcharge customers and use inferior products to increase margins. Salesmen, representing pipe, equipment, and service companies, regularly offered everything from cowboy boots to televisions to company men who could be influenced by this graft to send business their way. Pipe and wellheads were regularly stolen. Oil was siphoned off into water tanks, only to be picked up and sold by unscrupulous water haulers. A common saying of the day was that if the representative for oil purchasers didn't steal his own salary from the producer by underreporting oil on location, he wasn't doing his job. Producers underpaid royalties to landowners by applying adjustments and excessive charges. Deception was everywhere.

Just out of college, in early 1978, I went to work as a trainee for the Western Company, which provided well stimulation and cementing services. I was excited to move from the pipeline business—where I had worked for several years on corrosion-control systems and coatings while I went to night school—to the big time: drilling rigs in East Texas, one of the largest oil-producing regions in the country. I learned quickly, though, about the rules. One of the first things I was taught, besides how to handle an 18-wheeler, was how to fill out my Department of Transportation driver's log so I could work more hours than I was supposed to. I watched people drive huge well-servicing trucks on public high-ways with little to no sleep for days on end—and sometimes did the same myself. But that log was filled out right. During those days, I witnessed other dangerous practices and carelessness that were commonplace and saw several men severely injured and even killed as a result.

In 1981, I went to work for an independent oilman out of Dallas, who put me to work after, as a Western Company cementer, I had sat on a location for him on an East Texas Cotton Valley well. That well had stuck drill pipe and was kicking gas, trying to blow out. As a cementer, I worked on all kinds of jobs, from cement-ing casing in deep gas wells to repairing leaking pipe in some of the region's oldest oil wells. On this job, I worked for two straight weeks (in those days, relief was unheard of; once you were on a job, you stayed) pumping Black Magic, the chemical of choice to unstick pipe, and heavy mud to kill the well. We finally got the pipe unstuck, the well killed, and the casing cemented into the well. A few weeks later, the oilman's drilling and production manager called and offered me a job, and I jumped at it. I loved working for this operator. I went to blowout schools, attended open-hole well logging schools, and was a sponge for information, techniques, and, of course, crude oil field jokes.

The old superintendent I worked with taught me all about oil-production equipment—from downhole pumps to gas compres-sors—and in just a couple of years, I was running field operations for the whole company, driving from East Texas to New Mexico, drilling and working over wells. My experience has exposed me to drilling and production operations across the oil-producing regions of New Mexico, West Texas, East Texas, Louisiana, and the Gulf of Mexico. Years ago, I came to love the people (and food) in Deep South Louisiana, and I still go there every chance I get. Old toolpushers taught me to cook, and today my cuisine of choice when I'm doing the cooking is still rustic Cajun complete with Tabasco, cayenne, andouille, tasso, and a little lagniappe, "doncha know."

Over the early years, I have been part of a relatively small, tightly knit community to which change comes hard—except any change that brings more opportunities to drill, of course. In my early career, I witnessed practices that endangered lives as well as polluting our air, water, and the very ground where we live, work, and raise our children. I've also watched my industry deny that its activities have any effect on our environment, fight every effort to reduce those dangerous activities, yet take credit when improvements were forced upon it and worked.

In the last four decades, the United States has become dangerously dependent on foreign sources of oil, most of which are hostile, even as the oil and gas industry has encouraged burning more and more oil by supporting economic policies that squander, not conserve, supply. During these last 40 years, our elected leaders have been more than happy to kick the can down the road, taking campaign money from the industry and watering down or defeating every effort to establish a comprehensive energy policy.

The textbook example of the defense of status quo is the industry's steadfast opposition to fleet mileage standards in vehicles. Limits on gasoline mileage means less gasoline sold; they want lots of SUVs and gas-guzzling supercars to keep that demand high. As a society, we have helped drive a worldwide economy based on the burning of carbon-based fuels that release greenhouse gases and other noxious particulates into the air, but we have done little to advance beyond it. Even as we have failed to develop new sources of energy, our own supplies, especially of oil, have declined precipitously with few exceptions. Among those exceptions are the deep gas shale fields that can be successfully produced only with the now controversial massive fracturing treatments and the oil fields of the deepwater Gulf of Mexico. Though these two sources of energy extend our own supply, like every other conventional fossil fuel, they are finite. Even the most enthusiastic cheerleaders for the industry reluctantly admit that we have only about 50 years of domestic supply left. I personally believe it's much less than that, at least economically.

As a nation, we have been subject to embargos and price fixing from OPEC. Billions of dollars have been invested all over the world to develop reserves, and a good argument could be made that most of the violence and instability in the Middle East today is directly linked to multinational oil companies, like BP, and the governments with which they collude. To protect their monopolies, majors and independents in the oil industry have poured hundreds of millions of dollars into political candidates who carry their water, only to scramble, generally too late, to respond when political winds shift.

Which brings us to the subject of this book: the deadly blowout of BP's Mississippi Canyon Block 252 well, which caused the largest environmental catastrophe in the history of the United States. This is a tragedy that simply did not have to happen. It was caused by bad design, bad judgment, hurried operations, and a convoluted management structure. Add in silenced alarms and disabled safety systems, and the result was inevitable. But what were they doing out there to begin with? Why are we drilling in mile-deep water 50 miles out in the Gulf of Mexico? What has driven us to look for oil in such extreme environments, pushing on the edge of technology?

The disaster on the *Deepwater Horizon* that was drilling this well is a direct result of a number of complex failures, mostly human. The hubris of those who work in the oil industry always creates risk, even as the industry struggles to change its ways. Though much improved since I started more than 30 years ago, there is still that bit of a wink and a nod when it comes to changing actual behaviors. Even as companies like Transocean have safety observation programs, where employees are rewarded for pointing out unsafe practices, most of the results are somewhat superficial. They spend thousands of man-hours reporting slip-and-fall hazards and overhead dangers, but then they ignore, and even encourage, the disabling of entire safety systems. Indeed, Transocean has a policy that anyone on the rig can shut down operations if they deem an operation unsafe. At the same time, some employees can't name one time that anyone actually had the guts to do that. It looks good on paper, though, and sounds good in new-employee training classes. This dismissive behavior is embedded into the DNA of those who work in the industry, especially those who started back when I did. When combined with a truly threatening situation like that on the *Deepwater Horizon* on April 20, 2010, that behavior can erupt into a conflagration. Literally.

We all watched as the oil from the BP well came ashore weeks after the blowout, and we watched the now familiar videos of the oil in the surf and marshes along with the dead and dying birds. As tragic as this ecological catastrophe is, I can't help but think about the 11 men who were killed on the *Deepwater Horizon* and what killed them. As the disaster unfolded, BP's mantra, "All is going according to plan, but we won't know for another 48 hours," was disingenuous and did these men's memories dishonor while treating the general public as stupid. Clearly, despite all the assertions to the contrary, BP, and indeed the entire industry, was utterly unprepared to manage a blowout of this complexity and magnitude. They, and all the other companies that drill in the deepwater, assert in every permit filing that they are able to deal with potential blowouts and the ensuing spills that may occur, claiming that minimal environmental damage would result.

Nothing could be further from the truth. Before this blowout, my industry didn't have the slightest clue how to deal with a catastrophe of this magnitude, and many simply dismissed it, calling it "a black swan," or "one in a million," or saying that "accidents happen." Well, that's not good enough. As we watched BP struggle with this monster, saying every day that "this has never been done before," we all feared that it was the things we didn't understand that would probably be the most damaging. The unprecedented use of chemical dispersants at the seafloor was done on an experimental basis, even as the scientific community screamed for BP to stop carpet-bombing the vast spill site with dispersants. The damage is untold and likely to last for decades, but much of the media gleefully reported that the oil "disappeared" when dispersed oil stayed below the surface after the well's oil flow was finally stopped.

I have spent my career in this industry, immersed in its culture and practices. I've spent years in the field, on the finance and deal-making side, and managed companies large and small. I've watched the industry deny its negative impact on people's lives simply to protect profits, and I've listened as executives railed about how much money was being taken out of their wallets— instead of railing about the legacy they were leaving their own grandchildren. I've also heard oilmen describe members of their own families dying prematurely of cancer while denying that the air they breathe and water they drink may be causing that cancer.

As an insider, I believe I have a unique perspective on the oil and gas industry culture and how it contributed to the BP disaster. This book will examine the factors, both mechanical and human, that led up the blowout and subsequent oil spill. I will approach this story from my perspective as one in this industry, and I will attempt to describe the technical and mechanical issues in a language that hopefully everyone can readily grasp, assembling the events from investigative testimony and descriptions given by the survivors. Hopefully, this story can shed some light on what happened that fateful night, as well as sparking some conversation about where we are as a nation in our struggle for energy security, and how drilling in challenging environments fits into that struggle.

# | ACKNOWLEDGMENTS |

I simply could not have undertaken this project to write about the tragic events that began on April 20, 2010, without a publisher who *got it*, understanding not only the environmental catastrophe that we were witnessing, but also the human tragedies that were woven into every level of this complex story. Margo Baldwin of Chelsea Green Publishing certainly *got it*. When we first talked about creating this book, I didn't need to explain to her the importance of the story or to justify my approach. She just said yes, encouraging me to start yesterday to get it done.

Margo brought the entire Chelsea Green team to bear on this project, including the editorial director, Joni Praded, whose patience and calm voice kept me focused and moving forward, even when there was so much to write I couldn't get it out fast enough, at least in a readable form. As new events unfolded, even as we continued to work, Joni helped me incorporate the new information into the text. Patricia Stone, my project manager, Laura Jorstad, my copy editor, and Peter Holm, my designer, understanding that I was writing this story "live," allowed me the flexibility I needed to continue revising the text up until the last minute.

My very able researcher, Rachel Marcus, through her tireless efforts and long hours, helped me to tell this story by digging out the details of the events, even as they were happening. Her skills and dedication added depth and a richness to this story that I simply could not have achieved on my own.

The confidence that I could write this book came from the unquestioning support that I receive from my loving wife, Gracie, whose own literary accomplishments inspired me to begin writing myself. She encouraged me from the beginning, giving advice and gentle critiques as I developed my own style. When I began this project, she gave me the time and space I needed to get all this down on paper, even as it consumed my every waking moment. She created the environment that allowed me to focus completely on the story, putting the rest of our lives on hold while I wrote. For that, I am forever grateful and want her to know that I love her.

I continue to meet others from whom I learn new information and who provide for me a sounding board for my own thoughts and ideas. One of those I met after this tragedy occurred, known to the public only by his online moniker, Fishgrease, has become a close friend these last months. Our many late-night conversations about the Macondo well were very helpful to me as the crisis continued, and I found his insight enlightening, and his support unwavering.

Finally, I would be remiss if I didn't acknowledge those who gave me my start in the oil field lo those many years ago. Mike Abernathey was the experienced cementer who mentored me as a young trainee, teaching me the skills you can't learn from a book. His patient tutoring laid the foundation for my career. Bob Stallworth, Murray Helmers, and Carl Wheeler built on that foundation, giving me the opportunity to develop myself with hands-on experiences that gave me the confidence that comes only from doing. I will be always grateful to these men for giving me the chance to work in the business that I love, and I will never forget them.

Part One

# THE BLOWOUT

# | ONE |

# Unleashing the Beast and
# Paying the Price

Doug Brown found himself lying on his back—sprawled out on top of a metal deck panel on which, moments before, he had been standing, in the engine control room (ECR) of the *Deepwater Horizon* offshore rig. It was 9:50 p.m. and Doug, the offshore rig's chief mechanic, hearing an engine overspeeding, had just stood up from his computer to investigate when an explosion racked the otherwise normal night—the force of the blast hitting him from behind and knocking him face-first into the engine console. He bounced off the console and onto a deck panel covering a mass of electrical cables, but the panel collapsed and Brown fell through, ceiling insulation falling in on top of him. Looking out from his hole in the floor, he tried to orient himself. The room was dark and the hiss of escaping gas was deafening. Just as he tried to pull himself up and out, he was hammered with a second explosion—this one massive, driving him back down into the hole.

Dazed and in shock, he didn't notice that his leg was broken, or that he had serious lacerations, or that, of course, he had a concussion. As he again dragged himself out of the hole, the hissing transformed into a violent roar. Burning oil and gas took on a life of its own, intensifying into a firestorm, engulfing the drill floor, and incinerating the derrick that rose above the rig floor just outside the control room. The $CO_2$ fire-suppression system in the room had activated, displacing the oxygen. Breathing was difficult.

As Doug crawled out of the room and across the destroyed deck, he caught sight of Mike Williams, the chief electronics technician who had been working in his shop just adjacent to the engine control room. Bleeding profusely from a head wound, and also disoriented, Mike was crawling along the deck with a small flashlight clenched in his teeth, the only light to guide them out. Together they began making their way out of the blackness.

Having been just jarred out of bed, Steve Bertone, the rig's chief engineer, was running toward the bridge, dressing as he ran. The center staircase was blocked with debris, so he made his way to and up the port-forward spiral staircase until he reached the watertight door of the bridge. When he entered, the room was in chaos. Standing at his station, Captain Curt Kuchta was trying to make sense of what had just happened. Steve ran to his station, where his Simrad control screens confirmed his fears: The dynamic positioning system and thrusters that kept the floating rig in place were down. So were the engines. The rig was dead. He picked up the phone to call the engine control room, where Doug was supposed to be on duty. No dial tone. Setting down the useless phone, Steve rushed to the starboard window of the bridge, giving him a view of the rig floor behind him. The fire there was "derrick-leg-to-derrick-leg" and roaring over the crown, 240 feet above the deck.

At that moment, the port watertight door banged open. There stood Mike, so covered in blood that Steve didn't recognize him. Limping in behind him was Doug. "The engine room, ECR, and pump room are all gone!" Mike shouted, delirious from his head wound.

"What do you mean?" Steve asked.

"They're all gone. They've blown up," repeated Mike.

Steve ran to Mike, now recognizing his voice, and hollered out for medical supplies. "In the restroom!" someone shouted back. All Steve could find was a roll of toilet paper, which he applied to stop the bleeding on Mike's forehead.

Steve had not noticed Chris Pleasant, the subsea supervisor, who had come in while he was tending Mike's injures. Chris was standing before the blowout preventer panel. Just seconds before, he had declared to Curt that he was going to "EDS"—that is, activate the emergency disconnect system to get the rig away from the blown-out well.

"Calm down," Curt said, "we're not EDSing."

Noting what would later be described as the captain's deer-in-the-headlights hesitation, Chris ignored him and stepped to the panel anyway. Curt had already yelled at Andrea Fleytas, a dynamic positioning operator, who had taken it upon herself to call mayday on the marine radio without his authorization, even as the fire from the well reached up over the crown block of the derrick and all power went dead.

Don Vidrine, the well site leader who was called "the company man" by the old-schoolers, entered the bridge and joined Chris. Someone shouted across the bridge that they couldn't EDS without approval of the offshore installation manager—the top representative for Transocean, which owned the rig and leased it to BP. At that moment the manager, Jimmy Harrell, ran across the bridge shouting, "EDS! EDS!" Chris, looking at Don, said, "I'm gettin' off this well." Without hesitation Don said, "Get off." Chris reached out to the panel and pressed the EDS button.

After the EDS sequence was complete, Chris continued to stare at the blowout preventer (BOP) panel. Something was wrong. It had done exactly what it should have done when he pushed the EDS button, going through the sequence:

1. BOP stack to VENT
2. Blind shear rams CLOSE
3. Control pod receptacles RETRACT
4. Choke and kill line UNLATCH
5. LMRP UNLATCH

Chris just stood at the panel. Steve shouted, "Did you EDS? Did you EDS?" Chris shouted back, "Yes! I EDSed!"

But he hadn't. Once the sequence finished, the blind shear rams showed the red CLOSED light on the panel, as they should have. And the EDS showed complete. Yet instead of the rig lifting off and floating free of the blowing-out well, the roaring fire on the rig continued. Looking over at his hydraulic panel alongside the control panel, Chris realized that he had no hydraulic pressure. The flow meters showed no flow. The system had moved no fluid, closed no valves. The control pods at the blowout preventer 5,000 feet below had received no signal from the surface—or if they had, they had not responded. The pilot valves had not moved. Electronically, the EDS was successful. In actuality, nothing had happened.

The Beast was winning and would not let go, and the blowout preventer had just become a five-story hunk of useless scrap metal.

Fifty feet above the rig floor, Dale Burkeen, a veteran crane operator, so popular that he was called "big brother" by the younger hands on the crew, had been at his post at the starboard crane when the well blew. A bear of a man, Dale had tried to get clear of the crane and away from the well. As he ran from his cab, the oil and gas roaring from the well lit off right in front of him. The explosion lifted him from the catwalk and over the rail, smashing his body onto the crane pedestal and the deck below.

Moments earlier, when the first explosion happened, David Young, the chief mate, who also in charge of firefighting, had just left the bridge to get some rest in the crew quarters. He ran forward, back toward the bridge, and then immediately to the fire gear locker to suit up and pass out equipment as the fire team mustered. Except nobody mustered. As he was putting on his fire-resistant jacket, another crewman told him that someone was down on the deck below the starboard crane. He ran to the main deck, saw Dale's body, and, not knowing if he was dead or alive, realized that he would need assistance to move his sizable friend. Running back to the fire gear locker, he met up with Chris Choy, a young roustabout, who joined him to attempt a rescue. By the time they got back, the intensifying fire and debris had made it impossible to get across the deck. Reluctantly, they were forced to turn back. As they made their way back to relative safety, Choy stole a glance over this shoulder at the flames roaring over the top of the derrick; he then realized that there was no fighting this monster. Even if the firefighting system was working, it was no match for this kind of hell. Eventually, they joined the others on the bridge.

There the crew had realized that if they were going to have a fighting chance, they needed power. Any power. Curt and Steve decided to get the standby generator started. Steve took the radio offered by David, but they couldn't establish communications. Of course; like everything else that night on the rig, none of the radios worked, either. Steve decided to go anyway, with Mike insisting he not go alone. The motorman, Paul Mcinhart, who had also come to the bridge, went with them. As the three left the bridge and headed aft, Steve looked out onto the main deck; besides the roaring fire, he saw that the entire deck was covered with an inch of thick goo, which he later described as having the consistency of "snot." It was everywhere, and puzzled him. As they ran aft, they were able to see through the open blowout preventer storage room door into the moonpool, the huge opening below the rig floor, where pipe and tools ran between the rig floor and the well far below. Steve could see nothing in the moonpool but flames and black smoke. As the fire intensified, black smoke began circulating over the edge of the deck, only to be pulled back in by the firestorm to circulate again. Steve watched it return through the moonpool, rotating out over the decks again.

When they finally reached the standby generator room, the three went to work. They went through the start sequence and hit the button: nothing. After three or four tries, Mike checked the battery: 24 volts, good. Again, nothing. Steve went to the breaker panel, closed it, and then reopened it, hoping that would reset something. Still nothing. As Mike and Steve worked frantically, Paul went to the watertight door to look out. Steve felt the intense heat and told him to shut the door. Mike then shouted to switch to the second battery bank to see if that worked. Having done that, they hit the starter with hope. The generator was as dead as a hammer. Looking at each other, they abandoned that idea and ran back toward the bridge. It was clear that the EDS had failed.

On the bridge, the captain had already given the abandon-ship order, but Andrea and Yancy Keplinger, the other dynamic positioning operator, were still manning the marine radios. *"Abandon ship!"* Steve shouted. Andrea pushed the GDMSS button, which sent out a precise GPS location with an identifier message for rescue vessels, then headed out the starboard door and down the steps to the lifeboats. Chris, Don, and others had already abandoned ship and were in the number one lifeboat that was motoring away. The number two boat was long gone. The nine left on the rig grabbed a life raft, hung it from the davit, inflated it, and swung it out over the water after struggling with a line that Mike finally got cut with a pair of folding pliers he carried. Lying in a stretcher on the deck was Wyman Wheeler, one of the toolpushers, who was severely burned and had broken both his legs. David jumped into the raft, which was twisting in the wind 75 feet over the water, and pulled Wyman's stretcher inside. Steve followed him in carrying the other end. Andrea got in, along with Chad and Paul. Steve could feel the heat coming from under the rig and knew they didn't have much time. As the raft was being lowered, it began to swing and finally tilted 90 degrees in its sling, sending Wyman's stretcher and everyone else to the low side. As it got close to the water, smoke billowing out from under the rig filled the covered raft, nearly suffocating everyone inside. Andrea, terrified, finally screamed, "We're going to die!" as it slowly lowered at its crazy angle, threatening to dump everyone out.

After what seemed like an eternity, it touched down; Steve, Chad, and Paul jumped out, grabbing the raft to swim it out from under the rig, but it wouldn't move. Looking up, Steve realized the painter line was still tied off to the deck far above. No one in the raft had a knife, it being against company policy to carry one, and the knife that was supposed to be in the raft had gone missing. As they struggled with the painter, a pair of boots appeared out of the smoke above. It was Curt, who landed not 10 feet from the stranded life raft, having jumped 75 feet off the lifeboat deck. Moments later, Yancy landed a few feet from him. Looking up again, he saw someone run full speed across the helideck and jump. This was Mike, trying to aim his landing away from the life raft and others below.

Tragically, above them Jason Anderson, senior toolpusher, Dewey Revette, driller, and Stephen Curtis and Donald Clark, both assistant drillers, were all cremated unceremoniously by the well they had fought so long to tame. Floor hands Karl Kleppinger, Adam Weise, and Shane Roshto, with derrick man Wyatt Kemp, died together with two MI-Swaco mud engineers, Gordon Jones and Blair Manuel, in the mud and shaker rooms belowdecks in the first explosion. They never had a chance. Dale Burkeen had been killed, of course, when he fell from the starboard crane catwalk. The Beast had claimed all 11.

## Rescue of the Survivors

Meanwhile, the motor vessel *Damon Bankston* had been on station alongside the rig for a couple of days. In the hours leading up to the blast, Captain Alwin Landry had spent much of the day sitting comfortably on watch, monitoring the vessel's dynamic positioning screen as he watched his crewmen manage a mud-unloading operation. The *Bankston*, owned by Tidewater Marine, was in continuous service to the *Horizon*—its constant and very familiar companion. That morning, the *Bankston* had been taking on mud from the rig through a hose lowered from the rig's port crane. Since the floating *Horizon* was preparing for a move, Alwin was expecting to take on about 4,500 barrels of mud from the rig as it displaced its riser—which connected the well to the rig—and cleaned out its mud pits.

Anthony Gervasio, the *Bankston*'s relief chief, was down below filling the day tanks with fuel, preparing for the new day's operations, when he heard what he thought was a loud release of air or gas—he couldn't tell which, but it was loud enough to penetrate the hearing protectors that he always wore in the engine room. As he came topside, looking up toward the rig, he saw an explosion just aft of the derrick. The lights then went out, followed by a second, larger explosion that filled the derrick with flame. Recognizing that this was a really bad situation, Anthony didn't need an order from the captain to disconnect the mud hose from the rig. As he ran toward the hose, mud began raining down on the boat. With the assistance of a couple of crewmen, he managed to disconnect the hose and throw it over the side. Alwin was already maneuvering the boat away from the rig when the order came from the *Horizon*'s radio to stand off at 500 meters due to a well-control problem. The *Bankston* crew watched the fire as it reached over the crown of the derrick.

Having trained regularly on man-overboard drills and fast rescue craft (FRC) operations, Anthony was familiar with the procedure. He ran to the FRC, jumped in, and got it lowered to the water, away from the *Bankston*. Crewman Louis Langlois had joined him on the way down. Looking toward the rig, Anthony could already see flashes of light, which were reflective patches on workers' coveralls as they began jumping from the rig. Alwin, standing on the bridge, spotlighted the first survivor in the water so Anthony could see him. Anthony rushed out to the jumper, dragged him aboard, then went to the next spotlighted person. When the boat was full, they returned to the *Bankston*, where crewmen helped the survivors aboard. And so it went. As the flames intensified, Alwin maneuvered the *Bankston* to act as a shield from the heat and debris while survivors scrambled aboard on the starboard side. Back to the rig Anthony and Louis would go, circling and picking up survivors. At the same time, the lifeboats from the *Horizon* began arriving. As each pulled alongside, crew would assist everyone aboard.

When Anthony headed back toward the rig for yet another run, he spotted a life raft under the rig, but he saw someone in the water on the way. It was Mike. Dragging him aboard, he could hear Chad calling out for help from the life raft. Cautiously pulling up to the raft, he saw that it was still tied to the rig with the painter. Curt Kuchta, who was already in the water trying to get the line loose from the raft, swam over to the FRC, grabbed the knife Louis offered, swam back, and cut the painter. Anthony then tied them on and backed the raft down toward the *Bankston*. As the FRC pulled slowly toward the boat, a fishing vessel, the *Rambling Wreck*, arrived on scene. Curt sent word to them through the FRC to begin search patterns for any missing crew.

Coast Guard helicopters arrived shortly after, landing a rescue swimmer on the deck of the *Bankston* to assess injuries. The deck was so slick with mud from the blowout that the crew had to assist him to the cabin below where the injured were being cared for. Those in the worst condition were Wyman Wheeler, whom Steve had pulled onto the life raft, and Buddy Trahan, one of the Transocean executives who had come aboard with the BP visiting party that day and now had a broken leg and burns across his back. The swimmer made a quick evaluation and then began evacuating the injured from the deck by helicopter. The Coast Guard landed several other swimmers, and by six hours after the blowout all 16 injured, including Wyman and Buddy, were off the *Damon Bankston* and on their way to hospitals on the beach.

You'd think it was time to head home with the survivors. Oddly, though, under orders from BP, the *Bankston* made multiple stops on the way back, some to take on Coast Guard and BP personnel, some to drop off. The first stop on the way in was Diamond Offshore's *Ocean Endeavor*. Four BP/Transocean personnel were dropped off, and several medics were brought aboard. They then moved on to the *Matterhorn* platform and took on food, water, clothing, and more Coast Guard personnel. By the time they finally got back home at Port Fourchon, it was 1:27 a.m. on April 22. The crew was exhausted, injured, and still terrified; BP and Transocean decided that urinalysis would make the perfect welcome, followed by sequestration for the rest of the night in a nearby hotel. Before the workers were reunited with their families, release forms and statements were thrust in front of them. Some cooperated, some did not.[1-11]


## Now What?

On the night of April 20, 2010, 50 miles southeast of Venice, Louisiana, at approximately 9:50 in the evening, the oil and gas industry changed forever. Eleven good men lost their lives at the moment that BP's Mississippi Canyon Block 252 exploratory well violently blew out, shooting out the seawater that had been filling the riser connecting the semi-submersible rig to the well on the seafloor 5,000 feet below. The *Deepwater Horizon*, and its service vessel alongside, the *Damon Bankston*, had been pelted with a torrent of seawater and mud. Gas had followed almost instantly, sucked into the inductions of the two massive Wärtsilä deck engines that provided power to the rig and its systems. Breathing in the hydrocarbon-rich air into the fuel mixture, the overwhelmed number three engine, its emergency gas-sensing shutdown system disabled, ran away with itself, exploding in moments.

The blast of the exploding deck engine set off a chain reaction of events that ignited the giant jet of oil and gas coming up through the drilling rig floor, incinerating everything nearby, including the rig floor crew. The lights went out. Transocean's *Deepwater Horizon* had transformed into a latter-day *Titanic*.

That night, everything that could go wrong did go wrong. The deadman, a system that automatically closes the blowout preventer if it loses contact with the rig on the surface, failed. The emergency disconnect system, which separates the riser away from the top of the BOP, failed. The general alarm system was disarmed. Many of the gas sensors were inhibited, and emergency shutdowns for the engines were disabled or inoperative. Even the telephones and radios wouldn't work. The well was flowing uncontrolled to the surface while the rest of the crew scrambled for the lifeboats or jumped to safety unsupervised and unprotected. Once the survivors were rescued by the *Damon Bankston*, the Coast Guard and other workboats rushed to the scene but could only watch helplessly as the doomed rig began to drift and list, its dynamic positioning and flotation systems dead. Two days later, the rig sank, landing on the bottom 1,300 feet from the blowing-out well. It will remain on the bottom forever as not only a monument to those who died but also a stark reminder of the arrogance of overconfidence in sophisticated technology, and complacency bred from too many successes at shortcutting.

As the rig sank, the riser, still connected to the well, kinked and fell to the bottom, oil and gas roaring from several breaks in the pipe, putting even further stress on the BOP. As of this writing, oil gushed from the well for 87 days and the rig remains on the bottom.

The oil began washing up on beaches and wetlands of Louisiana's fragile coastline weeks after the disaster, and giant subsurface plumes of it, likely caused by unprecedented application of toxic dispersants on the seafloor, have now reached close to the loop current that could take it beyond the Gulf of Mexico. No serious person ever believed the fairy tale spun by the government, as BP stood mum, that the well was flowing at only the 1,000- and then 5,000-barrel-per-day rate that they stuck to for over five weeks; reluctantly the government now admits that the flow was as high as 80,000 barrels per day, not counting the natural gas being produced along with the oil. Even though the well was shut in on July 15, the world waited for the nightmare to end for five months, hoping this Beast would be permanently killed.

So. Now what? All of the shouts of *"Drill, baby, drill"* of the 2008 presidential campaign are silent. The free-market libertarians and Tea Party activists who called for deregulation of the oil industry and letting the "free market" solve our energy problems, and who also railed against a "government takeover of health care," ironically called for a government takeover of BP's cleanup effort at the height of the crisis. At the same time, though, their representatives actually apologized to BP, a foreign corporation, for this administration's requirement that it escrow $20 billion for the cleanup, calling it a "government shakedown" of private enterprise. As well, the same voices that speculated that the Obama administration was secretly planning to shut down the oil and gas industry, then criticized him for not going further in opening new areas to offshore exploration in early 2010, are now severely criticizing him for not doing enough to make BP get this well killed and making offshore drilling instantly safer. As they insist upon these oddly conflicting requirements, they also oppose any moratorium the president proposes to make deep-water drilling less risky. As usual, politicians from both sides are doing what they do best: playing politics. In the meantime, over 30 similarly designed rigs have been operating off US coastlines, and the world's first ever deep-sea oil plumes drift through the ocean causing untold damage. The scope of this disaster remains unimaginable—the impact unknown. I suppose we'll begin to get a sense of the total damage as species of sea life disappear from the Gulf and families with a multigeneration history in fishing go out of business.

In the midst of this catastrophe, real problems confront us. Some call for shutting down all offshore production. We simply can't do that, of course, since about 30 percent of our domestic energy supply comes from that region, and shutting it down would require us to import more oil, not less, from countries that hate us and use our own money against us.[12] Some call for shutting down only the deepwater. Same answer, since about 80 percent of offshore production comes from these relatively few deepwater wells.[13] Others call for alternative energy sources, including nuclear. These should certainly continue to be developed, but none of the alternative technologies is anywhere close to taking over any material part of our near-term energy demand, especially that used for transportation, and many issues remain unresolved, such as nuclear waste disposal. Answers are years away, if not decades, though, of course, we must start.

The disaster on the *Horizon* should be a wake-up call for all Americans, as it brings both energy security and responsibility for stewardship of our environment into clear focus. We all witnessed the largest environmental catastrophe in the history of the United States unfold before us on our televisions and computer screens while watching private enterprise, and government, incompetently struggle with the massive scale of the necessary response. This disaster has ignited a whole new debate about energy policy and controlling our own destiny, the role of multinational corporations in our economy, and our ability to cope with the effects of globalization on our country as a whole. As we struggle with the current issues, we still must face the future. This book will raise some of these important issues as we talk about how our national energy policy can (and should) be reshaped to address corporate culture, complacency, overconfidence, and negligence.

# TWO

# The Well from Hell:
# Drilling the Deepwater

The federal lease for Mississippi Canyon Block 252, located in 4,992 feet of water 50 miles southeast of Venice, Louisiana, was acquired by BP Exploration for just over $34 million in Minerals Management Service (MMS) Lease Sale 206 on March 19, 2008. Based on months of interpretive work on seismic surveys of the block, BP had targeted several large subsurface rock formations from the Miocene geological age, laid down millions of years ago, now some 13,000 feet below the ocean floor. Among the largest deepwater operators, BP felt confident enough in their interpretation of the seismic data that they were willing to drill a $100 million well, which they named Macondo, to test their prospect. This decision was the beginning of the story of "the Well from Hell."[1–3]

## The Anatomy of the *Horizon*

The ill-fated *Deepwater Horizon* was a fifth-generation semi-submersible drilling rig whose keel was laid on March 21, 2000, in Hyundai Heavy Industries Shipyard, Ulsan, South Korea, for the drilling company Reading & Bates Falcon. It was launched almost a year later, on February 15, 2001. The rig was a uniquely designed, two-pontoon floater whose columns slanted distinctly inward toward the deck from the waterline, giving it a very solid yet squat appearance. Almost 400 feet long and 300 feet wide with a 242-foot-high derrick set amidships, it was state of the art—a heavy-weather rig designed to work in the often harsh conditions of the deepwater. It was powered by six huge Wärtsilä diesel engines, each rated at 9,975 horsepower, driving six 7,000 kW (kilowatt) electrical generators. The deck could support 8,800 tons of pipe and equipment. The derrick—which carried the entire weight of the equipment that drilled deepwater wells—was capable of lifting 2,000 KIPs, or 2 million pounds. Its drawworks—essentially a huge winch that raises and lowers pipe into the well—were heave-compensated, meaning that as the rig moved up and down with the ocean swell, the hook holding the drill pipe in the hole stayed in the same place, maintaining the same weight upon the drill bit on the bottom of the hole.[4]

Historically, most injuries on drilling rigs happened on the rig floor, usually when rig hands were physically handling pipe. Traditionally, drill pipe was made up by wrapping a chain around it (called a spinning chain) and pulling it with the drawworks, screwing the pipe together; the drill pipe was then tightened, or torqued, with tongs, essentially massive pipe wrenches, also pulled by chains. Pipe and other tools were raised and lowered from the rig floor by catlines, controlled by wrapping rope around rotating drums (catheads) on the drawworks. On old-style drilling rigs, tripping pipe—running pipe into or pulling it out of the well—was, at least when going well, a dope-and-mud-covered ballet of perfectly timed movements of pipe, chain, tongs, and guys you wouldn't want to piss off in a bar. If something went wrong, though, people would lose fingers, teeth, and on a few occasions hanks of hair. Today it's much different, and generally safer. On a modern deepwater rig like the *Horizon*, virtually all of the handling, making up, and breaking out of pipe is done by robotic "iron roughnecks" and pipe-handling machines, making day-to-day operations much safer (and more efficient) than back in the day, since the floor hands don't have to dodge flying chains. In addition to the massive horsepower and equipment on the rig, the *Horizon* had the capacity to hold over 4,000 barrels of drilling mud, 13,000 barrels of drilling water, 300,000 gallons of drinking water, and 1 million gallons of fuel for the engines. A virtual floating town, it could sleep and feed 130 people who worked there around the clock.

Like all floating rigs—floaters—the *Deepwater Horizon* was essentially a ship. It had a bridge where the captain, or master, worked, except that this bridge was located not in the center of the vessel like on a normal ship, but port forward, or left front, under the helideck, which was under near-constant use as workers and parts to the rig were shuttled back and forth. The *Horizon* was self-propelled and utilized a dynamic positioning system driven by computer-controlled thrusters to keep it hovering precisely above the well, to which it was attached by the riser pipe. Dynamic positioning, or DP, is a marvel of modern technology that uses global positioning satellites and often a locator beacon on the seafloor to ensure that the rig stays in place under all but the most severe sea states. DP was once described to me as PFM—an oft-used oil field term meaning "pure fucking magic," and it really is. The system is remarkably accurate and reliable, not to mention critical to many vessels that work in the deepwater. Many rigs use the system, and most OSVs, or offshore supply vessels, also use it to stay on station next to a rig they are servicing. Some OSVs also utilize a fanbeam laser orientation system that keeps the vessel a certain distance away from the rig. The fanbeam system bounces a laser off the rig, feeding the DP computer with the distance by measuring the time it takes the laser to return. It works similarly to the laser guns that many police agencies now employ to catch speeders who use radar detectors to avoid getting caught. When the deepwater was first drilled, rigs were held in place by mooring systems composed of massive anchors and miles of heavy chain and cables, called in the industry wire rope. On a moored rig, computers control the tension on the mooring system to keep it centered over the well, adjusting for wind and waves. As floating rigs were designed for deeper water, the anchor mooring systems were replaced with dynamic positioning systems. Today most newly built deepwater rigs are constructed with DP to stay in place in ever-deeper water.

Floating rigs are classified according to the depth of water in which they can operate. These classifications, called genera-tions, group the rigs together in operating depth ranges, with fifth-generation rigs able to operate in up to 7,500 feet of water, while the new sixth-generation versions are equipped to regularly drill in depths of around 10,000 feet under harsh conditions. The *Deepwater Horizon* was designed for operations in 8,000 feet of water, even though it had operated deeper.[5]

## New Pressures on an Accomplished Crew

At the beginning of the last decade, the offshore drilling industry had become fragmented, with heavy price competition among the players struggling with highly volatile day rates. Small move-ments in future oil and gas prices had large effects on these rates as oil companies played drilling companies against one another to drive down costs. To combat this fragmentation and actu-ally reduce competition, a consolidation wave among drilling companies had begun a few years earlier, cresting in 2001, with the approval of the Bush administration, when Reading & Bates Falcon merged with Transocean, which had already merged with Sedco Forex. In that same year, Global Marine merged with Santa Fe International, forming GlobalSantaFe. Finally, in 2007, Transocean succeeded in eliminating other competitors by merging with GlobalSantaFe, forming the largest offshore drill-ing company in the world. During that massive consolidation, the *Deepwater Horizon*, now owned by Transocean, continued to work away, setting the world's deepwater record for a semi-submersible at 9,576 feet of water. Just prior to moving onto the Macondo well project, it had set another world record, this time for drilling the deepest oil well in history at 35,050 feet vertical depth in 4,130 feet of water.[6]

In accomplishing these feats, the *Horizon* crew also set a record of seven years with no lost-time incidents, one of the key measures of safety performance in the oil field. The industry prefers to call accidents "incidents," highlighting the fact that the vast majority of "incidents" occur as a result not of unsafe conditions, but rather of unsafe behaviors, and therefore are not accidents. Major incidents almost always involve multiple unsafe behaviors, sometimes combined with an unsafe condition. For several years when I was with El Paso Corporation, I was a member of the safety committee for the company. Our primary jobs were to review the safety performance of each business unit, oversee the investigations of incidents, and provide safety policy information and guidance to the business unit executives, managers, and employees. We would set corporate goals for safety performance, always shooting to reduce the incident rates to as close to zero as possible. The biggest challenge of managing a safety culture is keeping everyone focused. Maintaining a long-term perfect safety record is difficult, not just because of the law of averages, but also due to overconfidence and complacency on the part of executives, managers, and employees, which begins to pervade an organization that hasn't experienced a serious incident for a good length of time. When an organization goes for a year or more without a serious incident, a pat on the back is always in order. At the same time, training must be emphasized, because human nature is to let down your guard when you get comfortable, even in very hazardous environments like gas plants, pipelines, and drilling rigs. Certainly, with all their recent success, the crew on the *Deepwater Horizon* was experiencing both overconfidence and complacency as the rig moved onto Mississippi Canyon Block 252. The *Deepwater Horizon* was actually the second rig to drill on the Macondo location.

Deepwater wells, like all oil and gas wells, are drilled through a series of concentric pipes that get progressively smaller as the next-deeper string goes inside the last. There are several purposes for these strings of pipe. First and foremost, they keep the walls of the hole from falling in as the drill bit gets deeper, making it possible to drill at that depth in the first place.

Second, the pipe keeps the drilling fluid, or mud, inside the hole so it won't leak off into the substrata around the well. This is a very important requirement for drilling wells, especially to the depths seen today.

An important concept to understand as we talk about drilling is that oil and gas are not contained in large pools or caverns under the surface of the earth. They are contained in the pore space within the rock, trapped by impermeable barriers. To understand this concept of pore space, let's use sandstone as an example, as it is easiest to describe, and most clearly visualized relative to other types of reservoir rock. Sandstone is made up simply of grains of sand that have been subjected to extreme pressures and temperatures for millions of years as they are buried deeper and deeper over geological time. Even as the sand gets compacted, though, space still remains between its grains. Often pore space in sandstones is contaminated with clays or other substances, but pure sandstones are characterized by their large volumes of pore space. Underground, these pore spaces are filled with fluid. In most cases, that fluid is water. Shallow, onshore sandstones contain fresh water and are often the sources for water wells, but as you go deeper, that water becomes saltier. Hydrocarbons—oil and gas—are also contained in the pore space in sandstones. To stay in place, these hydrocarbons must be trapped by some kind of impermeable barrier, usually a layer of shale, or sometimes a seal formed by a fault (a movement of the rock millions of years ago). The trapped hydrocarbons are what we are looking for. As wells are drilled deeper, the pressure of the fluids in the rock, called pore pressure, increases gradually. Sometimes, though, that pressure is greater than would naturally occur, normally due to geological events that shifted the subsurface strata or because of the presence of impermeable barriers that trap fluids. Such zones are known as being abnormally pressured.

In order to control these downhole pressures, the driller must increase the weight, or density, of the drilling mud in the hole to raise the hydrostatic pressure at the bottom of the well and keep everything in balance. One challenge, though, is that if the driller weights up too much, the weight of the mud can actually overcome the strength of the rock below, with the fluids within it, and open fractures that will drain mud out of the hole. Losing mud into a formation lowers the hydrostatic pressure at the bottom of the hole, risking an influx of hydrocarbons into the well—and a blowout. Each geological formation has its own strength, so the drilling engineer must always estimate a balance between pore pressure and the frac gradient of the rock—the point at which the rock will fracture from the weight of the column of mud. In deep, complex wells, this balance can be very difficult to maintain. When the mud weight gets to the point where the column of mud overcomes the ability of the rock to hold back fluid, the well must be protected by running a string of casing and cementing it in place to relieve that pressure off the rock. As the well goes deeper and pressures increase, very close attention must be paid to the well as it talks back and communicates what it's doing or might do.

An old engineer taught me, years ago, that wells actually talk to you. You just have to understand the language they speak so that you can respond. Wells communicate in many ways, including how much gas is in the mud, whether you are gaining or losing mud, what solids you are getting back as you circulate the well, and how fast or slow it drills. These, plus other indicators, can help you stay ahead of a well to keep it from getting away from you. It's a 24/7 job that takes training, and years of experience, to understand.

## Early Warning Signs for the Macondo Well

BP's Macondo well was actually not started with the *Deepwater Horizon*. It was initially spud, or begun, on October 7, 2009, using Transocean's *Marianas* semi-submersible. An older rig, starting life in 1979, the *Marianas* actually uses an anchored mooring system rather than dynamic positioning, like the *Horizon*.[7] To start drilling the Macondo well, the *Marianas* first had to set a good foundation for the wellhead that was to support the blowout preventer and riser that would eventually be installed. The first process, once the rig was on station, was to set a heavy-walled pipe, 36 inches in diameter and about 300 feet long, on the end of drill pipe, and jet it into the seafloor—running seawater through a large drill bit on the bottom to help the pipe sink under its own weight as the water circulated through it. Over the next month, that pipe, as well as two more successive strings of pipe, were drilled and cemented down to about 3,900 feet below the mudline in preparation for landing the blowout preventer on the installed wellhead.

Operations on the *Marianas* continued to a depth of 4,023 feet below the mudline, or around 9,000 feet below sea level, when the rig was idled and evacuated for Hurricane Ida, which had formed the previous week in the Caribbean just south of Nicaragua. As the storm drove north, crossing over land, Ida entered the Gulf over the tip of the Yucatán Peninsula, continuing its northerly course and heading straight for Louisiana. Although it was downgraded from hurricane status each time it crossed land, Ida achieved that status a third time as it entered the Gulf on November 8, 2009. On November 9, it came across the Macondo location with winds of about 80 to 85 miles per hour, attaining wave heights of 10 to 15 feet. The rig, which had remained moored on location, sustained enough damage that it required a return to port for repairs. The Macondo location was temporarily abandoned, awaiting another rig. That rig was to be the *Deepwater Horizon*.

Operations on the Macondo well remained suspended until the *Deepwater Horizon* was moved in at the end of January 2010 to reenter the well and continue drilling operations. After going on dynamic positioning and rigging up, the crew splashed the blowout preventer on February 8, landing it on the wellhead in 5,067 feet of water and testing it the next day. By Minerals Management Service regulations, the blowout preventer must be tested every

14 days, and after every major operation. Each ram must be tested individually, using each of the two control pods to assure that all systems are working properly.[8]

When drilling in the deepwater, the riser pipe connects the blowout preventer, located on top of the wellhead, to the rig. To reduce the weight of the riser on the drilling rig, and to keep it from collapsing under its own weight, it is made buoyant by the use of flotation materials attached around the outside of the pipe. Also on the outside of the pipe are other smaller pipes, or lines, that give the rig the ability to monitor pressures in the well and to allow the well to be accessed while the blowout preventer is closed for safety. Risers come in sections as long as 90 feet and are handled by cranes and pipe-handling equipment as they are run through a rig's floor, out a large opening under the floor called the moonpool, and down into the water. The moonpool provides room for the rig crew to handle large tools, like the blowout preventer, and contains the riser tensioning system that keeps a constant pull on the riser as the floating rig moves up and down with the ocean swell. To run the riser to bottom, the blowout preventer is suspended in the moonpool and the first section of riser attached. That section, and other successive sections, are continued down into the water all the way to the seafloor, a process that takes about 24 hours in 5,000 feet of water. Once the blowout preventer is suspended over the wellhead on the seafloor, remotely operated vehicles (ROVs) guide the preventer to the latching collet on the wellhead using video feeds monitored at the surface by the crew. Once the blowout preventer is landed and latched up, the riser is hung in the tensioning system, the other lines are connected, and they are ready to drill ahead.

As the *Horizon* drilled deeper on the Macondo well over the ensuing three months, trouble began early and continued until that fateful day when the well got away from them with disastrous consequences. Even at shallow depths, the crew began experiencing multiple hole problems, gas kicks, and dangerous lost circulation zones—sometimes all at once. On four occasions prior to the 2010 blowout, the crew experienced well-control events. During one particular well-control problem in mid-March, the drill pipe became stuck, which means that it couldn't be moved in or out of the hole. Stuck pipe can be very dangerous during a well-control situation because it makes it much harder to control the circulation of the well and indicates poor hole conditions. After fighting the stuck pipe for a week, the crew finally separated from the stuck bottom-hole assembly, set a cement plug on top of it, and drilled a sidetrack hole, starting at 11,700 feet, to bypass the "fish" they had left in the hole. Any piece of pipe or equipment that is dropped or left in a well is called a fish. The attempt to retrieve a fish, not surprisingly, is called fishing. Fishing tool hands, or specialists, are some of the most valuable people in the oil field. They are generally gray-haired, they've been around for many years, and the good ones can save you millions of dollars if they are successful in catching their fish. In this particular case, BP chose to leave the fish and go around, making a new hole that gave them a better chance for success.

On it went. During testimony before the Joint Investigation Committee in May 2010, Mark Hafle, the BP drilling engineer who designed this particular drilling program, testified that the operation suffered "major lost circulation events" and "major well control events."[9] In an attempt to keep the well under control and manage the balance between frac gradient and pore pressure, BP applied for, and received, permission from the MMS to add contingency strings of pipe to make the well safer and allow the company to reach their objective depth of approximately 18,300 feet. As the days wore on, the well took its toll on the crew. After experiencing the multiple kicks, lost circulation, and stuck pipe, everyone was wary. During testimony before the Joint Investigation Committee in July 2010, Mike Williams noted that many of the crew had begun calling it the Well from Hell.[10]

## Faulty Design and Bad Decisions

Finally, after weeks of battling the well and one last lost circulation zone below 18,200 feet, the well reached its total depth. Logs were run—measurements made with electrical, sonic, and gamma ray tools run on wireline—to give engineers and scientists a picture of the subsurface intervals, their content, and their pressures. After the logs were analyzed, the decision was made to run pipe to the bottom to prepare the well for completion and production. It was at this point that BP likely made one of its worst errors—an error that likely played a key role in the blowout of the Macondo well. BP engineers decided to run one *long string* of casing from the bottom of the well all the way to the wellhead. This meant that the only protections provided against a flow of oil and gas into the wellbore were the cement that would be pumped down the casing and up around the outside of the very bottom, the mud above the cement, and the seal assembly that would be locked in right at the wellhead on the ocean floor. If the cement job failed to hold back the oil and gas, it could easily channel its way up through 13,000 feet of mud all the way to the wellhead. If the casing hanger seal assembly failed, a release of well pressure directly into the riser was a real possibility. Another route of failure could be the float equipment inside the casing, allowing gas to channel up the inside of the casing string. A safer design would have been what's called a liner, which is a string of pipe that only lines the open hole up into the last string of casing. At that point, another seal assembly is set and cemented in the top of the liner that provides an additional barrier to the flow of oil and gas. BP decided against that design for unknown reasons. Perhaps it was money; I doubt that. I believe it was time pressure and the desire to get pipe on bottom before something else went wrong. For whatever reason, BP engineers ultimately recommended the long string design, and their managers approved it.

In the deepwater, since the wellhead is on the seafloor, everything is done long-distance. For example, when casing is run to the bottom of the well, it only goes back to the wellhead on the seafloor, not back above water level, as is the case on wells in shallower water. To get the casing all the way to the bottom of the well, drill pipe is attached to the casing and forms the connection from the wellhead, through the riser, to the rig on the surface. Cement is then pumped down the drill pipe, through the casing, and up the outside of the casing. Between the bottom of the drill pipe and the casing, a setting tool is used to attach the casing inside the wellhead. Once the casing is cemented, the drill pipe is removed, leaving the casing behind.

Some have said that several decisions made in designing the Macondo casing could have contributed to the blowout. First, BP chose a lightweight, nitrified cement, which is essentially formed by injecting nitrogen into the slurry as it's pumped. This cement is generally used on shallower strings of pipe where lost circulation and water flow are significant problems; it's not typically used as a completion cement deep in the well. Additionally, BP decided to run only 6 centralizers—devices that keep the pipe centered in the hole—rather than the 21 recommended by Halliburton, the cementing service company on the rig. Another factor could have been the size of the pipe relative to the size of the hole that was drilled. The very last section of hole was drilled to a diameter of only 8½ inches. The pipe they ran was 7 inches in diameter, leaving very little room for a cement sheath around the outside, especially since they were using a lightweight slurry. Another contributing factor could have been the lack of circulation prior to cementing the pipe. Rather than circulating mud "bottoms up" to make sure that the gas was out of the well and the mud was clean of cuttings from the drill bit and debris, the crew circulated for only a short while before they began cementing.

The Halliburton cementer reported good returns of mud back to the pits during the cement job, though some question that report. Good returns indicate full circulation, meaning no mud is being lost in the well, allowing the cement to rise to the level on the outside of the pipe that was intended. Lost circulation indicates a problem, such as cement leaking off into one of the lost circulation zones that had plagued this well. After the cement job, the crew set the seal assembly in the casing head to seal off the outside of the casing they had just run. The driller on duty at the time, Micah Burgess, pulled tension on drill pipe to get the casing hanger setting tool to shear out or release from the casing hanger and seal assembly; he then came out of the hole with the drill pipe to lay down the setting tool. Once the casing is set and cemented, the crew normally prepares the well for temporary abandonment and gets ready to mobilize the rig to go to the next job.

Typically in the deepwater, most successful discoveries are abandoned for a period of time to allow production facilities to be built, delivered, and set on the well location. The respite also allows the operator to finalize a development plan, including additional wells that may need to be drilled. To make a well safe for abandonment, a cement plug is usually set below the mudline; then all seals are tested again before the riser is displaced with seawater prior to pulling the blowout preventer back to surface. Before unlatching the preventer, seawater is always displaced into the riser to prevent the heavier drilling mud from entering the ocean environment during the unlatching. Usually, however, several operations must be completed before displacement occurs. First, a lockdown sleeve is run into the casing hanger seal assembly to lock it into the casing head and prevent it from moving. Then the top cement plug is set.

In this particular case, however, BP apparently decided to shortcut these procedures by displacing the riser with seawater first, then performing the other operations. Several witnesses to the morning meeting that day reported that there was disagreement between a BP company man and the Transocean representatives about one of the tests to be performed and the procedure for displacing the riser afterward. Witnesses reported that the BP company man was Well Team Leader Bob Kaluza, and the Transocean representatives were Jimmy Harrell, the offshore installation manager, Dewey Revett, the driller, and Randy Ezell, a senior toolpusher. Doug Brown, the chief mechanic, was present at the meeting and reported the confrontation as a "scrimmage," but Harrell later denied there was any argument. According to descriptions, Kaluza overruled the Transocean representatives, and apparently they capitulated. Before the top plug could be set and the riser displaced, several operations had to be performed, including two tests to be sure the casing and its seals were holding. The two tests were relatively straightforward. The first, a positive test, applies pressure down into the casing and seal assembly to make sure they are holding. To do this, Burgess, who was still on duty, ran a tubing/drill pipe work string in the hole, stopping just above the blowout preventer. He closed the blind shear rams on the BOP, sealing the well from the surface, and had Halliburton use their pumps to pressure up on the casing through the kill line—one of the lines that runs down the outside of the riser into the bottom of the BOP. Burgess testified, and the drilling report shows, that he held 2,500 psi on the casing for 30 minutes, and it held successfully. After this positive test was run, Burgess opened the top ram and the crew ran the drill pipe/tubing string down to 8,367 feet, about 3,300 feet below the mudline, where the top plug had been approved by the MMS to be set.

The second test for well integrity is very important—the negative test. This test looks for leaks coming into the well from the outside, like checking a boat for leaks when you put it into the water. It is performed by relieving pressure from the well to see if there is any flow into it. When I was a cementer, I ran negative tests on different kinds of cement jobs all the time to make sure the well we had just been cementing was holding and not flowing. The facts about the negative test on the *Horizon* that day are difficult to glean because the drilling report doesn't show the procedure that was actually followed, everyone who was involved at the time of the negative tests died in the blowout, and survivors have given conflicting accounts. We'll never really know what happened, but we have evidence that the negative tests failed and that the well was indeed flowing into the riser.

A normal negative test attempts to put the well in a neutral condition, so you can tell if anything is flowing from it. This is complicated in the deepwater, since you can't see the wellhead, a mile below you on the seafloor. In most tests, seawater is pumped down the drill pipe and up the casing to just above the BOP. Usually, a set of pipe rams or an annular preventer is closed around the pipe and the well is then monitored for any flow. According to Transocean's preliminary report and witness statements, a heavy spacer fluid containing lost circulation material, or a plugging agent, was pumped ahead of the seawater, but not pumped far enough to get it above the BOP. Later accounts asserted that the annular preventer actually leaked the spacer back into the blowout preventer. In any case, because the heavy spacer was in the BOP when pressures were checked, it could have easily caused abnormal pressure readings, and apparently did. The crew got confusing readings on both the drill pipe and kill line; since the annular preventer had leaked, they had also lost mud, causing additional concern. They decided to run another negative test.

## The Final Night

Distractions during times like these can be dangerous because they take everyone's minds off the job at hand. The night of April 20 was unusual, filled with distractions. First, everyone was likely beginning to relax now that this very difficult well had finally been cased and cemented. Even though cementing a well can upset its balance and make it dangerous, most crews breathe a sigh of relief when that plug bumps on the cement job indicating that cement is all the way to bottom, and when the well appears static. This was very likely the case that evening. The negative testing procedure came right at the time of a short change of the tour (pronounced *tower*), meaning that a new crew was coming on at 5:30 in the evening, rather than 11:30, the normal tour change. Short changes often come when crews are changing places, day crew to night and vice versa. It happened that this change occurred smack in the middle of the negative test. Another distraction was that visiting executives from BP and Transocean were touring the rig at the time. A meeting was going to be held about an hour later, in which the crew was going to be recognized by the visitors for their seven-year safety record. These visits, no matter how important, always distract employees from their primary duties, and no doubt this was the case on April 20.

After much discussion on the rig floor, Jason Anderson, the senior toolpusher who had just come on tour, decided to run another negative test, with the agreement of BP representatives. Witnesses reported that this second negative test was successful, even though the pressure readings were still abnormal, and Anderson ordered the crew to move ahead with displacing the mud and spacer out of the riser with seawater in preparation for Halliburton cementers to set the top plug. From here the accounts vary widely. One is from Christopher Haire, the Halliburton cementer on duty that night. In a television interview with KPRC investigative reporter Robert Arnold in July 2010, Haire recalled a very different sequence of events surrounding the negative tests. He claimed that both tests were failures, adding that during each test, he opened his tanks to the kill line on the well and it flowed water into his tanks—indicating that the well was flowing. After shutting in the kill line after the second test, Haire reported what he saw to the rig floor. He waited an hour with no communication from the rig crew and then walked up the staircase to the rig floor to ask what was going on. Haire reported to Arnold that four Transocean employees stated that they had performed a third negative test from the rig floor, and everything was okay. They instructed him to prepare to run the top cement plug after they displaced the riser.

The annular preventer used for the test was opened, and the mud and spacer were pumped out of the well with seawater. Randy Ezell, the senior toolpusher whom Anderson replaced that night, called the rig floor about 9:20 one last time before going to bed for a final check. "I've got this," he reported Anderson to say. Ezell went to his room and was soon in bed with the lights off. A few minutes later, after a call to his wife, his phone rang. Ezell checked the clock: 9:50 p.m. The voice on the other end was Assistant Driller Steve Curtis. "We have a situation. The well is blowing out and mud is going to the crown." Ezell was horrified. "Is it shut in?" he blurted. Curtis said, "Jason is shutting it in now. We need your help." Those were the last words Ezell heard from the rig floor. The explosions happened before he could get his boots on. He spent the rest of his time rescuing Wyman Wheeler and Buddy Trahan, both severely injured in the blast.

In the hours leading up to the disaster, the Well from Hell was screaming at the crew that it was going to blow out, but nobody could understand the language it was speaking. Serious mistakes were made, not only that night on the rig, but on multiple occasions leading up to it. The *Horizon*'s crew, one of the most experienced in the deepwater, their supervisors, their managers, and their executives had all been deafened by their many successes. This rig had just drilled the deepest oil well ever. It also held the semi-submersible record for water depth. It had just set a seven-year record for safety. Their operator, BP, could do no wrong. They had been so successful that these wells were becoming rote, commonplace. Complacency and overconfidence had set in.

In the weeks prior to the blowout, some of the crew on the *Horizon* felt a foreboding and even experienced premonitions that something was going to happen. Natalie Roshto, testifying before the Joint Investigation in July, talked about her husband, Shane, a floor hand killed in the blowout. She spoke of his pride in his work in the offshore and how he was contributing to his country's energy security. His last time at home, though, he was different; he spoke to her of the kicks and mud losses they had experienced. He discussed his worry about the safety on the rig and the pressure to get this well finished. When he talked with her by telephone for the last time that day, he continued to speak about the problems they were having. Natalie had always been afraid of the helicopter rides Shane took to the rig. She always felt that he was safe once he got there. It's heartbreaking that her confidence was misplaced. Shelley Anderson, Jason's wife, said that on his last hitch home her husband was obsessed that everything be in order. He even drew up a will and showed her how to handle things herself, including how to operate their motor home. Before he left that last time for the rig, Jason told her that if anything happened, Transocean would take care of her. As of this writing, she's still waiting.

Next to that of manned spaceflight, the technology used to drill the deepwater is some of the most sophisticated ever imagined by man. Drilling wells in 5,000 to 10,000 feet of water has been likened to working on the surface of the moon or heart surgery on the bottom of the ocean. Those analogies are accurate, but I often think of deepwater drilling as like driving a car from the backseat; you can reach the steering wheel, but it's hard to control and you can't get your foot on the brake pedal very easily. Because the distance between man and well is so far in the deepwater, the technology must be the link between those two. As we've been made painfully aware, when the technology fails, or people fail, the consequences are catastrophic.

As much as we want to celebrate our technological leaps into the deepwater, the old driller's and cementer's skills are still just as important. Drilling is drilling. Cementing is cementing. Well control is well control. The deepwater just emphasizes the critical need for those skills in our offshore engineers, scientists, drillers, and service personnel because the margin for error is reduced to close to zero. If nothing else, we will hopefully learn these lessons from this disaster.[11–20]

# | THREE |

# How Blowout Preventers Don't Work

The magnitude of the disaster in the Gulf of Mexico after well control was lost was a result of the spectacular failure of the blowout preventer to shut in the well. Usually considered a fail-safe device, it was anything but when it came time to use it. It sat useless on top of the well until it was finally raised to the surface on September 4, 2010. Because this was the key piece of evidence in the ongoing government and legal investigations, the US Department of Justice took control of the massive device as soon as it landed on the deck of the *Q4000*, the semi-submersible construction platform that raised it. Since that day, the government cut the live public video feed it had provided on the deck of the *Q4000* and has given little information about the BOP's whereabouts, or their intentions for it, except that it was shipped to the NASA facility at New Orleans for analysis.

The blowout preventer is the defense of last resort in any kind of drilling, especially in the deepwater. It is designed to control a well that is trying to flow to the surface—hence the name, blowout *preventer*. While some of the BOP's functions are routine—allowing fluid to be circulated in the well, for instance—it plays a critical role in sealing the well and, when emergencies arise, severing the casing or the drill pipe so that the well can be disconnected from the riser and the connected rig. To keep from needing to use the BOP in disaster mode, proper balance in the well must be maintained so that the weight of the drilling fluid—the hydrostatic pressure of the column—is sufficient to hold back any influx of hydrocarbons in the well (often referred to as kick), but not so heavy as to overcome the strength of the rock that holds it up.

The BOP's primary protective device is the ram preventer, essentially a large valve that closes horizontally over the well to seal it. Modern-day subsea blowout preventers often contain up to six sets of ram preventers of various configurations as well as annular preventers. An annular preventer works a bit like a giant tire inner tube. Located on top of the ram preventers, it inflates with hydraulic fluid, squeezing around the drill pipe and sealing off the void outside the pipe. Annulars are generally designed for lower pressures than ram preventers.

The first ram-type BOP was designed and built (in Houston, of course) in 1922 by oilman Jim Abercrombie and machinist Harry Cameron to overcome the common problem of the early oil field gushers. In those days, well control was unheard of, and the only (and obviously dangerous) indication of a successful well was a blowout that covered the countryside, and all the workers, with oil. As wells went ever deeper, and pressures got higher, blowouts became larger and more deadly, not to mention expensive. A better way of blowout prevention had to be developed.

Abercrombie, who had started as a roustabout in the Goose Creek field about 20 miles southeast of Houston, eventually came to own several drilling rigs, operating in South Texas and along the Gulf Coast. Always needing repairs done on his rigs, Abercrombie soon met Houston-based machinist Harry Cameron. They eventually formed a partnership, calling it Cameron Iron Works, and together designed the first successful ram-type BOP.

Called the Cameron MO BOP, their design quickly became the industry standard for well control. Back then, the BOP contained only one set of rams designed simply to close around drill pipe and seal off the well. It was manually screwed shut with socket wrenches that extended from under the rig floor to allow a safer location to close them. Even in those early days, the MO BOP was tested to 3,000 psi, a technological marvel for the time. Though Harry Cameron died in 1928, just a few years after their invention, and Abercrombie in 1975, their company, now called Cameron International, is the world's largest manufacturer of

BOPs, and it is an industry leader in pushing the technological envelope for high-pressure deepwater well control.[1]

## The Modern-Day Tool of Last Resort

Even though the modern BOP resembles its ancestor in shape, it is light-years beyond in function and operation—when it works. When it doesn't, you may as well have an old MO BOP on top of the well, or a lug wrench. They would both be about as useful. The BOP *not* working is a much bigger problem than the oil and gas industry would have you believe.

Let's talk about subsea BOPs such as the one that sat on top of the blown-out BP well like so much scrap iron. Despite the massive failure of this particular stack, subsea BOPs are a marvel of modern technology, but they are complex and subject to various single-point failures, in which one valve is controlled by multiple systems. The risk of single-point failure is the Achilles' heel of the BOP.

In 2009, a risk management organization, Det Norske Varitas (DNV), was commissioned to do a confidential study for Transocean on subsea BOP reliability, using a database of 15,000 wells drilled in North American waters and in the North Sea from 1980 to 2006. It found 11 cases of blowouts in deepwater wells where the BOP was required to be activated. Yet only in six cases were the BOPs successful in shutting in the wells and avoiding oil spills in the surrounding water. DNV classified that failure rate at 45 percent.[2] An earlier study, performed by West Engineering for the MMS in 2004, showed that of 14 recent newly built deepwater rigs, only 7 even tested their subsea BOP's ability to shear drill pipe. Of the seven BOPs tested, four failed to cut the pipe. West's conclusions were prophetic:

> This grim snapshot illustrates the lack of preparedness in the industry to shear and seal a well with the last line of defense against a blowout.[3]

A subsea BOP can weigh as much as half a million pounds. It's as tall as a five-story building and consists of large rams, with a bore that is wider across than the steering wheel in your car. This assembly is designed to withstand very high working pressures and the maximum working pressure of the rams in the *Deepwater Horizon* stack is 15,000 psi. Cameron International built this BOP about 10 years ago, at the same time the rig was built. However, it was subsequently modified, with some of those changes documented, some not—and in at least one case, making it less safe. For example, in 2004, Transocean agreed to remove the variable-bore blocks from the bottom ram and invert it, converting it into a test ram. This test ram would allow for faster BOP tests, but at the expense of redundancy and safety. In a letter to BP, Transocean noted the increased risk, and it got BP to agree to pay for additional downtime expense should the BOP need to be pulled for repair due to the lack of backup rams.[4]

Starting from the bottom, the stack that covered the Macondo consists of a hydraulically controlled connector that latches on to the wellhead, allowing the rig crew to set and pull the stack remotely with hydraulic controls. Above the connector are the five ram preventers themselves, the core of the BOP. On top of the ram preventers is another assembly, called the lower marine riser package. The LMRP is equipped with an emergency disconnect system, which allows the rig to separate from the blowout preventer in the event of an uncontrolled blowout or other emergency. Activating the EDS saves both the rig and the BOP from extreme stresses and prevents oil from gushing up from the well. Also in the *Horizon*'s LMRP are two annular preventers, designed to close around whatever pipe is in the hole during well flow or testing. The failure of the EDS to separate the rig from the BOP was one of the critical failures of the safety systems during the blowout of the Macondo well.

The ram cavities in the stack contain different types of closing devices, or ram blocks, that can close around different sizes of pipe (called variable-bore rams) and close off the hole. In the event of a critical well-control situation, the ultimate fail-safe cutting the drill pipe and sealing off the well is called the shear ram or blind shear; it's designed to cut the drill pipe that may be in the hole. It can also seal the hole if there is no drill pipe in the well when the well begins to flow. The big daddy of all rams is called the super shear, designed to actually cut casing while it is being run into the hole. As we now know from the BP disaster, the weakness of shear rams is that they can't cut *all* pipe that may be in the hole, especially the new high-strength pipe used today. They are also not usually designed to cut the various running tools or drill pipe tool joints that are much heavier than the pipe itself. The BOP stack is usually designed so that, in a well-control situation, the chance of a drill pipe tool joint being in the blind shears is minimized—but this assumes that the floor crew is on the floor and has time to go through the shear and shut-in sequence. In the case of the *Deepwater Horizon*, we now know that the rig floor crew was trying to close in the well when the gas flowing to the surface lit off, killing them all.

The annular preventer, the high-strength rubber element that inflates around the pipe and holds back well pressure, is designed to actually allow pipe movement, including tool joints, unlike ram preventers that lock around the body of the pipe. This ability to move pipe through the annular preventer allows rig crews to "strip" or push pipe into the hole under pressure, to get it deeper into the hole in the event that the well kicks while they don't have pipe all the way to the bottom, or to keep the pipe moving to keep it from getting stuck, a common problem when a well is trying to flow.

The BOP has valves in the ram cavities to which the kill line and choke line are attached. These high-pressure lines, which run from the BOP to the surface, allow the well to be flowed from or pumped into under high pressure, even though the BOP rams are closed. It is these lines that BP used during their "top kill" operation to attempt to pump mud into the well, and then later used to flow oil to the surface for containment. Similarly, the choke and kill line valves on the last capping stack were also planned to be used for flowing the well to containment vessels, if that had happened.

The BOP is activated, or fired, with explosive force applied from what are called accumulator bottles. These bottles, which operate at 3,000 to 5,000 psi, contain pressurized hydraulic fluid, which is used to close all the rams on the stack. The control system is also hydraulic and tells the BOP what to do by applying pressure to various valves in the control pods. The BOP is activated when the operator opens a pilot valve from the surface. That signals the control pod to open a control valve, releasing accumulator pressure to fire the rams, causing them to close with explosive force—as much as 5,000 psi.

Redundancy is key to BOP control, so there are multiple panels, accumulators, and control pods. On a subsea stack, there are two control pods, always called the yellow pod and blue pod, that generally sit on the base plate of the LMRP, designated by color to make it clear which is which when maintaining or activating them. The yellow control pod is the one that famously leaked in the weeks prior to the blowout of the Macondo well.

On deepwater rigs such as the *Horizon*, these pods are controlled and tested by sending electrical signals from the rig floor down an umbilical line alongside the riser. Several control panels that operate the BOP are placed in strategic areas of the rig to allow quick access by the floor crew, facility managers, or the subsea supervisor. Additionally, the BOP can be operated by remotely operated vehicles, which can access intervention panels on the stack itself. A BOP must also have what's called a deadman—an automatic system that activates the BOP should it lose communication with the rig on the surface. In US waters, only an umbilical deadman is required—one physically connected to the surface. Other countries mandate an acoustic system that communicates with sound signals transmitted through the water between the stack and the rig. Many acoustic signals per second are transmitted and received between these components. If the BOP loses this acoustic contact with the rig, it automatically activates the BOP, shutting in the well on its own.

These acoustic systems cost about $500,000 each, so the MMS chose not to require this device on rigs operating in federal waters. Would an acoustic system have saved the *Deepwater Horizon*? Probably not, since both deadman systems, umbilical and acoustic, require a blowout preventer that actually works. Since the BOP itself failed, it didn't matter whether the deadman was wired or acoustic.

## What Went Wrong on the Macondo

There are a number of reasons that a BOP can fail to close. Subsea BOPs are complex, precision devices that operate under extreme conditions often controlled from the surface while in up to 10,000 feet of water. The units are subject to corrosion, hydrostatic pressure, high internal pressures, and near-freezing temperatures. Stresses from movement in the riser, the floating rig, and ocean currents are common. Because of these conditions, strict regulations govern the use, testing, and inspection of these units. As previously noted, the BOP must be tested every 14 days according to federal regulations, as well as after every major operation such as setting casing. The yellow pod and blue pod must be tested alternately. If any test fails, the drilling contractor is required to discontinue drilling operations until repairs are made.

According to testimony in the Joint Investigation and witness reports, either BP or Transocean failed to follow these regulations.

During his testimony to the Joint Investigation, Transocean's subsea supervisor, Christopher Pleasant, testified that he had overseen the preventive maintenance of the *Horizon*'s BOP stack before it was splashed—as they say in the industry—onto the Macondo well. He said that it tested properly and had no leaks. However, he also testified that, at the time of the blowout, he attempted to close both shear rams and separate the rig from the well with the EDS. He activated the blind shear rams from the control panel on the rig's bridge, and he reported that, even though the panel electronically went through the appropriate sequence, the hydraulic gauges showed no fluid movement. When he activated the EDS, he got the same result. He abandoned ship with the rest of the crew a few minutes later after the BOP failed to operate. We know the rest of that story.[5]

In June 2010, Tyrone Benton, an ROV technician with subcontractor Oceaneering, gave an interview to CNN about a leak he found in the BOP hydraulic system. He told CNN that months before the blowout, he had detected the hydraulic leak on the yellow control pod on the *Deepwater Horizon* BOP during an ROV-conducted inspection. ROV inspections of the BOP occur regularly, since the ROV is the eyes of the drilling contractor in the deepwater. In the interview, Benton asserted that he had reported the leak to his bosses at Oceaneering, who reported it to BP and Transocean. He also claimed that, rather than suspending operations, then pulling and repairing the pod, BP merely disabled it, eliminating half of the BOP's redundancy. BP, of course, immediately refuted Benton's claim.[6] However, it's well known that, after the rig sank, BP indeed pulled the yellow pod with ROVs and made repairs, later acknowledging a leak in that pod.[7]

In testimony to the Joint Investigation on July 20, 2010, Ronald Sepulvado, a BP company man on the *Deepwater Horizon* who had been relieved four days prior to the blowout to attend a training course, testified that the yellow pod did indeed have a hydraulic leak, and that he had reported the leak to his team leader in Houston. He also testified that he "assumed" that the MMS had been notified. Investigators noted that nothing was mentioned about the leak on the daily drilling report that was sent to the agency on Sepulvado's watch. Sepulvado testified that he thought the leak was minor and didn't affect their ability to drill, even though the MMS regulations clearly state that if a component is not operable, drilling operations must be discontinued until that component is repaired and put back into service. Robert Kaluza, the BP company man who had relieved Sepulvado and was on duty at the time of the blowout, took the Fifth twice, and at the time of this writing he has so far refused to cooperate at all with the Deepwater Horizon Joint Investigation.[8]

Soon after the blowout, Transocean and BP said the BOP had been tested and was operable with no leaks. Now, after multiple reports saying otherwise, a BP official has admitted, under oath, that indeed the yellow pod was leaking, but that BP and Transocean had continued operations anyway. After the blowout, the BOP *and* EDS completely failed.

We now know that not only the BOP was modified, but so was the ROV panel on the stack. In testimony to the Joint Investigation on August 25, 2010, BP official Harry Thierens testified that they struggled to shut the well in with ROVs for a day before realizing that the panel had been reconfigured at some point, but the changes were left undocumented. As they tried to close the middle pipe rams around the drill pipe in the BOP, the rams they were actually firing were the tests rams that had been installed in 2004. These rams were useless because they couldn't close around pipe or hold pressure from within the well. Thierens also testified that when Transocean brought drawings of the BOP originally provided from Cameron, they contained modifications, hand-drawn in red marker. The undocumented modifications certainly delayed the attempted shut-in—but worse, they could have caused the BOP failure.[9]

Long after the blowout and sinking of the rig, BP was indeed finally able to trigger all six sets of rams on the BOP after making repairs to the yellow pod and the plumbing on the stack, but the action did not successfully shut in the well. BP conceded, however, that erosion to the seals and damage prevented it from sealing. Rams are designed to slam shut, not be flowed through for extended periods of time, especially at the very high rates and pressures experienced during a blowout. Erosion of the ram faces and the bore was by then likely severe, further reducing the BOP's chances of ever actually shutting in the well and sealing it.

We also eventually learned that there were at least two pieces of drill pipe stuck in the stack. At the time of the blowout, BP was in the process of temporarily abandoning the well after cementing and had drill pipe in the hole to set the last cement plug. When the well blew, the drill pipe probably got hung up in the BOP, possibly because at least one ram partially closed at some point during the chaos. Later, possibly as late as when the rig actually sank, the drill pipe in the riser above the BOP likely parted, a piece of it falling alongside the existing drill pipe inside the stack. It took until July for BP to finally acknowledge the presence of the second piece of pipe and make provisions for dealing with it when they set the capping stack. This second piece of pipe is also what likely caused the wire saw to jam during the attempt to cut off the riser in June.

As an aside, we know that BP executives from Houston were visiting the rig on the fateful day, ironically, to celebrate seven years of no lost-time accidents on the *Deepwater Horizon* and tour the facility. According to witness accounts, they had toured the rig and had been on the bridge using the dynamic positioning simulator minutes before the blowout. I can't help wondering if, during their visit, something was inadvertently done on the bridge that could have neutralized the BOP control system. I can't think of what that would be, but the question lingers about why every single safety system on the rig failed. And the coincidence that guests had been on the bridge manipulating controls only minutes before the BOP failed nags at me.

As I write this, we don't know yet why the BOP failed. Did it close on a tool joint or the casing hanger? Were there two pieces of drill pipe stuck in the rams when they tried to close? Was there not enough accumulator pressure to fire the blind shears? Or did it fail to close at all? Why did the EDS fail to separate? Was the battery in the deadman low? Were the umbilicals severed in the blast? These are questions that we can answer only now that we have the damaged stack on the surface for study. There's a lot to learn in this tragedy.

Which brings us to the next step: How do we improve on the BOP and the procedures for its use before we go back to drilling in the deepwater?

## Time to Start Taking Safety More Seriously

On May 27, the Obama administration announced a six-month moratorium on drilling in the deepwater.[10] In June, that order was stayed by a federal judge in New Orleans, and the case continues to wind its way through the courts, casting doubts on the near-term future of deepwater drilling.[11] During this pause, there are many things the industry could be doing to be ready to resume safely when they have the chance. Rather than complaining about the suspension, and making excuses that this disaster was a "black swan" (I love that one) and that deepwater drilling is safe the way it is, the industry really should be leading the way in developing new technology, devices, and procedures to make it truly safer.

To understand why, we need look no further than the evidence provided by the "augmented tests" on the two blowout preventer stacks on Transocean's *Development Driller II* and *Development Driller III* prior to splashing them on the relief wells that were drilled in response to the BP blowout. During these tests, newly required by the Bureau of Ocean Energy Management, Regulation and Enforcement (BOEMRE), these two BOPs failed four of the tests: Both the EDS and the deadman system failed due to bad valves, and the casing shear rams failed to close due to a faulty control pod. These failures occurred on fairly new BOP stacks, built in 2004 and 2005.[12]

So what does this mean? *Transocean was preparing to splash faulty BOPs for blowout prevention on relief wells being drilled to kill a blowout caused by a faulty BOP.* Jesus. I find it pretty chilling that Transocean, had it not been for the new tests, would have run these state-of-the-art BOPs that suffered virtually identical failures to that of the *Deepwater Horizon's* stack.

Besides implementing new testing and maintenance regulation, improvements must be made in stacks to allow ROVs to intervene much more readily than they could on the *Horizon* BOP. After the blowout, and the realization that the damaged BOP was simply not going to work, precious weeks were wasted as ROVs painstakingly cut away the choke and kill lines and fabricated new valves to provide entry into the BOP. Since an accident like this was never supposed to happen—an accident in which the rig, riser, and all control systems were lost—no one had even contemplated making a provision for accessing the blowout preventer from anyplace but the rig. There was no valve that could be accessed easily to allow pumping into, or flowing out of, the well using another vessel. Adding these valves, I believe, is a relatively simple fix that has to be engineered and deployed as part of a broader external intervention program in the event, God forbid, that this kind of disaster ever happens again.

Accidents always provide a catalyst for improvements. That's been a well-learned lesson during a century of aviation experience, with every major catastrophe leading to improved safety, improved performance, and advanced technology. As an example, worldwide availability of GPS navigation has virtually eliminated lost airplanes, one of the most common problems in aviation as recently as 20 years ago. Those same GPS units help pilots to manage their fuel consumption, since running out of fuel is probably the most common cause of aircraft accidents.

This disaster, as gigantic as it is, can do the same for energy policy and industry safety as airplane crashes have done for aviation safety—if the government, the politicians, *and* the industry embrace the lessons learned as an opportunity to improve the way we do things, rather than exploit it for political gain. However, given Americans' short attention span and the industry's concern for the money in its wallet today, I'm not holding my breath for any mature, well-thought-out strategy. Bureaucrats will make a lot of noise, then hide when the hard decisions need to be made. If we stick to our usual MO, politicians will call for short-term remedies that make for good sound bites but improve no policies. And the industry will complain about reduced profits and threaten to move all of their deepwater rigs to other countries, firing all of their American workers instead of fixing their problems.

Since the BOP is the last line of defense, this is where the attention needs to be paid. Clearly, well casing design, drilling plans, and safety programs are extremely important. However, there is one piece of equipment that simply can't fail when it comes time to use it: the BOP. We can figure out, and indeed we already know much of what went wrong on the *Deepwater Horizon*. Poor judgment, complacency, and impatience mixed with a highly modified, partially disabled, and uninspected BOP made for a volatile combination. The last line of defense failed, and that just can't happen again.

# | FOUR |

# Technology Reaching Beyond
# Our Ability to Control It

For a successful technology, reality must take precedence
over public relations, for Nature cannot be fooled.
— RICHARD P. FEYNMAN

If there is one thing we've learned from this catastrophe, it is
that the technology that has been developed to drill the deepwater is at once marvelous and terrifying. As we sat for hours
glued to televisions and computers, watching remotely operated
vehicles navigate the seafloor at extreme pressures and temperatures, we witnessed what to some would seem like magic—robotic
machines with mechanical arms that can operate with surprising
dexterity or crushing strength. These "underwater roughnecks"
have a camera for eyes and joysticks for brains, operated by very
skilled "pilots" sitting on the surface watching the action on their
own screens. Most people who were seeing ROVs for the first
time were shocked that they even existed. Little did they know
that ROVs were developed almost 50 years ago and have been in
common use in the deepwater for well over 15 years.[1]

Like ROVs, the technology of the deepwater has expanded the
regions in which we can search for oil and gas. Once thought
impossible, operating in 5,000 to 10,000 feet of water is now
commonplace. Rigs, the riser systems that connect the wells to
the rigs, well-control systems (whose reliability is now in question), subsea production systems and pipelines—all of these exist
and form a spiderweb of infrastructure that spans from the barrier
islands of South Texas to the white sand beaches of Alabama.

And it's not only in the deepwater Gulf of Mexico. These technologies are applied to reserves in deepwater all over the world, including the coasts of Africa and Norway, the shores of Southeast Asia, and Arctic regions. These sources will continue to fuel our demand for hydrocarbons in this worldwide economy, but the challenge is that our technical and scientific knowledge seems to always go just beyond our capabilities to control it if something goes wrong.

It has always been so.

## Lessons from Sea and Space

The tragedy of the *Titanic* is a textbook example. Built almost 100 years ago, it was a marvel of its time. Huge, capable of carrying over 3,500 souls, and fast at 24 knots, the ship was thought to be virtually unsinkable, though that term was never used publicly until after she sank. The night of April 14, 1912, was moonless, and the sea flat calm. Because of these conditions, the ship's only warning system, the lookouts, could not see icebergs or hear waves breaking against them as they usually could. Technology had reached beyond Captain E. J. Smith's ability to control it when he piloted his ship at over 20 knots into the Atlantic ice field, dooming the ship. At this speed, the ship was overrunning its ability to see, and the captain was flying blind, likely under pressure to make an early arrival in New York on the ship's maiden voyage.

The gigantic *Titanic*, almost 900 feet long, contained 16 main watertight compartments below the waterline, with doors that could be electrically closed from the bridge; she could stay afloat with four of them flooded. Because the iceberg was hit with a glancing blow on the starboard side, though, five compartments were opened to the sea—one more than she was designed to handle. The ship was doomed and went down by the head less than three hours after the collision. One of the most visible shortcomings of the ship was her lack of lifeboats. Though she could carry over 3,500 passengers and crew, she carried lifeboats for fewer than 1,200. There was absolute confidence in the ship's ability to stay afloat should something happen to it, and because the shipping lanes were so busy (it was believed), passing ships could lend assistance. One of the suspected reasons behind the dearth of lifeboats was to give more room for people to walk on the boat deck.

One common theme in these kinds of accidents is that the laws of the day couldn't keep pace with the growth of technology. In the Titanic's case, the law of the day required only 16 lifeboats for ships over 10,000 tons. When the law was written in 1894, ships the size of the Titanic had not been contemplated. The ship actually had more than was minimally required, carrying an additional four collapsible boats, though the addition made almost no difference in the needless loss of life, since so many of the boats were launched only partially full.[2]

Philip A. S. Franklin, vice president of the White Star Line, said the day after she sank, "I thought her unsinkable and I based my opinion on the best expert advice available. I do not understand it." Many lessons were learned in the sinking of the Titanic, and technologies in ship design, metallurgy, navigation, communication, and safety took huge leaps, but at the tragic cost of over 1,500 lives.[3]

Similarly, aviation technology advanced quickly as it was being developed. In the early days of heavier-than-air flying, successful flights were commonly concluded with a spectacular crash. In 1911, the notion of commercially flying mail took hold, launching regularly scheduled US Airmail service. Eventually that service was expanded coast-to-coast, and brave (and just a little crazy) pilots would fly in all weather to get the mail to its intended destination. Air navigation, adapted from maritime skills, was by dead reckoning and visual reference to landmarks. Eventually, light beacons were placed on towers along routes that pilots followed at night. The life span of an airmail pilot was pretty short, though, and the mail often didn't get through on time due to weather, mechanical failures, and of course crashes. The demand for scheduled services helped lead to development of navigation instruments and stabler, faster aircraft, and it also expanded our ability to fly in adverse conditions. As with the *Titanic*, though, the technology that first allowed planes to fly in bad weather, at night, exceeded a pilot's ability to actually see beyond his aircraft's own range. Technology had to catch up.

Spaceflight grew technology by leaps and bounds, always reaching beyond our control if something went terribly wrong, such as when an oxygen tank on *Apollo 13* exploded as the ship headed for the moon in 1970. The now famous story of how the aeronautical engineers got the crew back safely has been the subject of several books, including *Lost Moon* by the spacecraft's commander, Jim Lovell. Renamed *Apollo 13*, that book became the basis for the movie by the same name made in 1995, directed by Ron Howard and starring Tom Hanks.

On the night of April 13, 1970, an overheated oxygen tank on the spacecraft exploded when bare wires connecting to a fan, designed to stir the oxygen in the tank, sparked, setting the tank insulation on fire and causing the oxygen in the tank to boil off faster than it could vent, building up extreme pressure. The wires were bare because insulation had been melted off when the tank overheated to as much as 1,000 degrees during a previous ground test, but nobody had noticed. The tank, it turns out, contained an older thermostat designed for a lower voltage that had not been upgraded when NASA changed its specifications, a fact that also went unnoticed. Consequently, the lower-voltage thermostat had melted shut, keeping the tank heater continuously on rather than cycling on and off as designed. With the error undetected, the faulty tank remained installed in the service module and was refilled with oxygen when the time came for the *Apollo 13* flight. In space, when it exploded, it blew out the side of the service module, which contained the astronauts' oxygen, water, and, of course, the rocket engine that was supposed to get them home after their walk on the moon. The explosion damaged all but one oxygen tank, as well as the fuel cells. Because the oxygen was used for both breathing air and electricity in the fuel cells, it was the lifeblood of the spacecraft; had the lunar module—the vehicle designed to detach from the larger spacecraft, land the astronauts on the moon, and return them to the craft—not been there for the astronauts to use as a lifeboat, they would have all surely died.

Returning *Apollo 13* safely to earth was a huge accomplishment, from both a leadership and an engineering perspective. Gene Kranz, the NASA flight director, refused to allow the team on the ground to give up on getting the three astronauts, Jim Lovell, Fred Haise, the lunar module pilot, and Jack Swigert, the command module pilot, safely home. Over just a few days, they were able to devise a plan that allowed the lunar module to be used as a lifeboat, with the astronauts breathing its air and using its batteries for power, since they had to completely shut down the command module. Afraid to use the service module engine, not knowing whether or not it was damaged, they used the lunar module engine to propel them home after rounding the moon. However, the lunar module was not designed for longtime human occupation, and its carbon dioxide scrubbers were soon exhausted. Of course, no one had ever thought that this kind of situation would present itself, so the *square* command module scrubbers, which were not being used, would not fit where the exhausted *round* lunar module scrubbers fit. Engineers on the ground devised an assembly that used what the astronauts had on hand to build a carbon dioxide scrubber out of the square canisters. It worked.

On April 17, after four harrowing days, the *Apollo 13* command module reentered the earth's atmosphere and safely splashed down in the Pacific Ocean, only 4 miles from the aircraft carrier USS *Iwo Jima*. Bringing the three astronauts home safely is still regarded as one of NASA's greatest achievements, even though the mission itself was a failure. The electrical fault caused by the out-of-spec thermostat illustrates how small failures can cause catastrophes beyond the control of anyone, even those who design these sophisticated technologies.[4-7]

The difference between these historical examples of tragedies, or near tragedies, that have occurred and the Macondo well blow-out is the degree of consequence. Generally, the results of disasters like the *Titanic*, aircraft crashes, and spaceflight failures are contained. Even though 1,500 people died on the *Titanic*, that was it. There was no further damage, and the maritime industry learned a lot from that tragedy. Disasters such as BP's well blow-out, on the other hand, have almost overwhelming consequences: Not only did 11 men die, but thousands of square miles of open ocean were covered in oil, habitats were destroyed, thousands of animals were injured or killed, many cleanup workers and other people living near the spill experienced health problems, and industries that depend on the Gulf, like tourism and fishing, were crippled. The oil industry, along with the nuclear and chemical industries, is unique in that the price for technological failure can be paid for by millions of people, costing billions of dollars.

## When Extreme Technology Collides with Human Error

As we've talked about in previous chapters, deepwater drilling technology has grown rapidly as water depth and weather conditions have become more extreme. The rigs, drilling systems, and well-control systems have grown more sophisticated—yet not more reliable, at least not when it really counts. It's not that drilling is completely unreliable, though. It's more that it's unpredictable, requiring more redundancy than is now legally mandated, which is now the subject of much discussion among the regulators at the newly minted Bureau of Ocean Energy Management, Regulation and Enforcement.

Deepwater drilling, and the technology it takes to produce it, didn't draw much public attention until April 20, 2010. The vast majority of Americans are more than happy to fill up their SUV with relatively cheap gasoline, preferring not to think about where it comes from or how long the supply is going to last. It is one of those rare products that everyone buys, but until there is an environmental disaster, few of us see or think about it unless the price goes over $3 a gallon. It travels from the well to the refinery as oil; then it moves, as gasoline, from the refinery to a pipeline or truck, to the gas station tank, into your car, and finally into the air as exhaust while it propels you wherever you want to go. With vapor recovery units now common in major metropolitan areas, you don't even smell it when you're filling up your car, but it's there, nonetheless, a lot of it coming from deepwater production in US waters.

Until the Macondo well blowout, deepwater technology and safety were scrutinized by only a few, and the cozy relationship that had developed between the regulator, then the MMS, and the operator rang few alarm bells as complacency bred from years of success set in. This catastrophe that we are facing now, though, was foreshadowed in some pretty notable failures in deepwater operations over the last decade, some caused by mechanical failures, some by computer, and most led by or exacerbated by human error as control was lost. Let's take a look at a few examples.

BP's Thunder Horse field in the Gulf of Mexico's Mississippi Canyon is only about 40 miles from the Macondo well. Discovered in 1999 in Block 778, drilled by Transocean's now famous *Discoverer Enterprise*, it sits in 6,000 feet of water and contains up to a reported billion barrels of oil from about 25 wells producing from "wet" trees, meaning subsea. The project has been plagued with numerous problems mainly from water depth, temperatures, pressures, and volumes flowing from the wells. BP says they have to develop 100 new first-time technologies to make this project work.[8] Ironically, the prospect was originally named Crazy Horse, for the rebellious Lakota chief, but BP changed the name at the request of his descendants.[9]

On May 21, 2003, a Thunder Horse development well, Mississippi Canyon Block 822 Number 6 well, suffered a parted riser when the *Enterprise*, on dynamic positioning, was pulled off station by heaving and/or currents. The rig had just started to come out of the hole when the riser parted with a huge *bang*, since it had over 2 million pounds of tension pulled in it. The riser parted in two places, one right above the lower marine riser package, looking eerily like the Macondo well after the riser was cut off, and another at about 3,200 feet. Fortunately, the deadman function worked on this BOP, automatically cutting the drill pipe and shutting in the well. The synthetic drilling mud in the riser, though, was unloaded into the Gulf when it parted. One of the close calls in this incident was that part of the riser fell alongside the BOP, narrowly missing its control lines and ultimately ending up leaning against the stack.[10]

The problems for Thunder Horse continued when its massive floating platform, designed to process up to 250,000 barrels of oil per day, almost sank when one of its pontoons flooded in high seas as Hurricane Dennis passed through the Gulf in 2005. The platform was saved and repaired, but it could have been a total loss. After repairs to the platform and damaged subsea manifolds, the field finally went on production in 2008, nine years after its discovery.[11]

Another notable riser separation happened in 2005 when Ensco 7500, a dynamically positioned semi-submersible rig drilling for Anadarko, was pushed off station as a tropical weather system moved into the DeSoto Canyon area of Green Canyon Block 252 in over 4,000 feet of water. One of the thrusters on the rig had been down for maintenance, and the remaining thrusters and engines could not keep it on station. The crew was actually pulling out of the hole when weather kicked up sooner than they had expected with 10- to 14-foot seas. As they drifted farther off station, the variable-bore rams were closed and the drill pipe set down on them. The reason this is done is to assure that the drill pipe tool joint, which is much heavier than the pipe itself, is spaced properly above the blind shear so it can be cut if necessary. As the crew desperately tried to displace the riser to seawater, the winds rose to 61 knots, pushing the rig 380 feet off the well. The offshore installation manager gave the order to emergency disconnect, but the angle was so high that the EDS function wouldn't activate, plus they were holding pressure of 3,000 psi on the rams, which some said caused the delay in the EDS sequence. The subsea engineer switched control pods and EDSed again. After four minutes, the lower marine riser package finally unlatched and lifted off the blowout preventer after the blind shear rams cut the drill pipe. The tension was so strong by the time it released that the riser slip joint, right under the rig floor in the moonpool, recoiled, damaging the tension ring. No hydrocarbons from the well were leaked, but the crew did end up dumping about 700 barrels of synthetic drilling mud into the Gulf. The rig drifted 3 miles, trailing 4,300 feet of riser and the LMRP under it, before it was brought back under control.

Obviously, the crew was caught flat-footed by a tropical depression that rapidly developed into a Category I hurricane, dubbed Cindy. When it crossed the Yucatán Peninsula, it was traveling at up to 17 knots, not giving the rig enough time to prepare for its arrival. That, along with the out-of-service thruster, put the rig, and the crew, outside the envelope where they could control the outcome. They were very fortunate that the dangling riser didn't hit another subsea structure or run aground.[12]

Besides weather, depth of the water, currents, and mechanical failure, the risk of human error in keeping up with technology growth in drilling is significant—but of course that's true not just for this industry. It's everywhere. We are inundated with information coming from smaller and smaller devices that are increasingly complex. I regularly use three or four devices myself, from an iPhone to an iPad, a laptop, and a desktop computer. Each serves a set of important functions and improves my productivity, but rather than simplifying life, they make it all the more complex. The cockpit of a modern private aircraft is a perfect example of the danger of technologically overloading a pilot. The glass panels in aircraft today, just like dynamic positioning, are stellar examples of the PFM we've talked about before, but you have to understand that magic by being trained in it, familiar with the menus, the functions, and the displays—yet always prepared to revert to the old steam-style gauges underneath the fancy magic screens should they go blank at exactly the wrong moment, which they are wont to do. Don't get me wrong; when everything is working properly, you're ahead of the aircraft and comfortable with what's going on, they are amazing devices that keep you spatially oriented and give clear situational awareness. However, if you are in weather, something is acting up, the radio is busy, and some air traffic controller sitting in a dark radar room is vectoring you around, you can reach overload in a matter of seconds. And those seconds, depending on how well you are trained and practiced, can be critical. All that PFM becomes pure fucking hell *right quick* if you don't instinctively know which button to push or knob to turn to get you back on track when the time is right.

Deep drilling, whether onshore or offshore, is like that in spades, with offshore itself certainly getting really complicated *right quick*, as in the Ensco 7500 incident we just talked about. Like in flying, the human factor is key and the human–machine interface must be clearly understood and accounted for to ensure safe operations. Complacency and success are the biggest enemies. The perfect example of that was the *Deepwater Horizon* offshore installation manager inhibiting the general alarm, putting more reliance on the *human* in the human–machine interface. That decision didn't consider the time when the humans in the human–machine interface couldn't get to the alarm as things really went wrong.

Miscalculations like this happen frequently in the oil field, just as they did on the Macondo well blowout. Some incidents are downright ironic, like the one in February 2000 on Diamond Offshore's *Ocean Concord*, a semi-submersible drilling for Murphy Oil on Mississippi Canyon Block 538. The Minerals Management Service had just come out with a Notice to Lessees, or NTL, alerting operators to the problem of accidental disconnect of marine risers, proposing that guards be installed on BOP panels to prevent crew from accidentally pushing the LMRP disconnect button. Murphy notified Diamond, and Diamond management ordered that the new guards covering the disconnect button be installed immediately. At the time, they were running a 9 ⅞-inch liner into the well, so the subsea engineer, who happened to be new to the job, was ordered by the offshore installation manager to install the guards. He went to work, installing the guard on the BOP panel in the manager's office. He didn't disable the panel, not knowing how, and, while drilling holes to install the guard to keep from inadvertently pushing the button, he . . . wait for it . . . inadvertently pushed the button. Newer panels now require a two-button push-and-hold procedure, but this one was an older model where you just pushed the button once. The LMRP unlatched, dumping over 800 barrels of synthetic mud into the environment. Losing hydrostatic pressure, the well started flowing into the water, unloading about 200 barrels of oil into the Gulf before they finally got the BOP shut in.[13]

The new subsea engineer was clearly out of his league, and poorly trained. Indeed, the offshore installation manager admitted that this was only his second hitch on the *Concord* as subsea engineer. On its face the incident seems pretty stupid, but, as is common in the oil field, many operations personnel receive primarily on-the-job training. This is true even today. On the *Deepwater Horizon*, the subsea superintendent and subsea supervisors were trained on the job, receiving little outside training beyond well-control school. They learned BOPs and their systems on their off hours on the rig, competing for these coveted jobs. Training this way, while providing valuable experience in actual conditions, risks wide gaps in both knowledge and skills—as has become painfully apparent during the investigations into the cause of the Macondo well blowout and subsequent BOP failures.

Dynamic positioning, one of the most sophisticated technologies utilized in the deepwater, is very reliable, but even when working properly it's especially subject to human error; the costs are huge when a mistake is made. An example of serious operator error, combined with an equipment malfunction, demonstrates this weakness. In 2007, GlobalSantaFe's *C. R. Luigs*, a dynamically positioned drillship, was completing a well at Green Canyon Block 652 in over 4,000 feet of water when the dynamic positioning operator was doing normal preventive testing. As part of the test, the operator transferred location data to the secondary console, started it, and then shut down the primary console. Satisfied the test was successful, he restarted the primary console and transferred the location data back, not noticing that the data was somehow corrupted during the transfer. With the wrong data, the dynamic positioning system thought the ship was in the wrong location (it wasn't) and began pushing the rig out of its watch circle, or safe operating area, which the dynamic positioning operator and captain immediately recognized. They restarted the secondary console but inadvertently transferred the corrupted data there, too, so the rig continued to chase the wrong location. The operator was working so fast inputting corrections that he locked up the dynamic positioning computer. Once maximum riser angle was reached, the positioning system initiated an EDS, which was completed at the driller's panel, separating the lower marine riser package from the blowout preventer. The well was safely shut in, but they ended up dumping 500 barrels of heavy completion brine out of the riser into the Gulf.[14]

There are many other examples of how badly things can go when technology grows beyond our ability to control it. It has become very clear, watching the Macondo well blowout story unfold, that there is very little (or no) margin for error in the deepwater, and bad situations can escalate very quickly into catastrophes.

Technology can be a servant, but it can also be a master if we don't keep up. In the deepwater, hiring, training, and keeping good people is key. The offshore is tough work and not for everyone. Many people, especially the well educated and highly trained, choose jobs that are easier fits with family life, offer regular hours, and aren't so physically demanding. Because of this, the human resources on these very sophisticated facilities can be thin, leading to having people in jobs of authority who may not be quite ready for it. Also, the need to keep the rigs manned leads to delaying or cutting short some training, creating knowledge gaps in sometimes critical areas.

Combine the human challenges with sophisticated equipment that needs frequent and heavy maintenance, the weather, the seas, and geological challenges *on top of* the corporate culture and continuous pressure for profitability, and you have the perfect recipe for disaster. We have certainly witnessed the results.

Part Two

# THE RESPONSE

# | FIVE |

# A Rogue from the Start: Influencing World Politics, Gambling with Safety

B P. Its vital statistics say it all. Fourth largest company on the Global Fortune 500[1] and third largest oil and gas producer in the world, just behind Royal Dutch Shell and Exxon Mobil.[2] Largest oil and gas producer in the US, with 2009 revenues of $181 billion (that's about half a billion a day). Reserves of 18.3 billion barrels equivalent. Owns interest in 16 refineries, has operations in 30 countries, has 80,000 employees.[3]

It is, by every measure, a Supermajor. Yet this is the company that has now presided over the largest environmental catastrophe in the history of the United States, and it seems that they have learned little from this disaster or previous ones, except for how to protect their public image. And they've been lacking even in that. At this writing BP has shown little propensity for transparency, and in fact they conceal information about their operations while laboring to create the illusion of openness. Based on their history, this is no surprise.

In recent years, since its takeover of Amoco and Arco in 1998 and 2000, BP has been plagued with safety and environmental violations and tragic accidents reflective of their cost-cutting culture and laser-sharp focus on profits even while they tout their "green" image with a cute daisy logo and slogan of "Beyond Petroleum." This year, though, many have invented new slogans that may be more fitting, such as "Beyond Pollution," or "Beyond Preposterous," or "Beyond Prosecution." One that I think fits particularly well is "Beyond Propaganda," for their penchant for downplaying negatives with happy talk and the dismissive "everything is going according to plan" mantra that was so often repeated in the early days after the Gulf blowout.

BP's lack of transparency has been genetically encoded into the company for over a century of exploiting resources and the countries within which it operates. To understand BP's corporate culture, we have to look all the way back to its origins. The company's influence is surprisingly interwoven with modern world history—to the point that we can say BP has actually influenced world events, with shattering consequences.

BP started out life as an investment by Englishman William d'Arcy, who had made his fortune in Australian land and mining after moving there with his family at a young age. In 1901, shortly after returning to Britain, d'Arcy funded a search for oil in Persia, which is now Iran, buying a 60-year lease from the Iranian monarchy for almost the entire country for £20,000 (and a few well-placed bribes) and a 16 percent interest in the oil revenues paid to the monarchy. He gambled his entire fortune funding an exploration team led by explorer George Reynolds, who searched the country for oil. Having drained his coffers by about £500,000, in 1904 d'Arcy sold much of his interests to another British company, Burmah Oil, for an additional investment of £100,000. Finally, in May 1908, they struck oil, discovering one of the Middle East's largest fields, and d'Arcy's group owned it all.

Anglo-Persian Oil Company was formed, with d'Arcy on the board of directors. A big supporter of converting the British fleet from coal to oil was First Lord of the Admiralty Winston Churchill, who later said of the discovery, "Fortune brought us a prize from fairyland beyond our wildest dreams." Soon after, the British government bought the company and built the world's largest refinery in Abadan, on the Persian Gulf coast, to process the oil and ship it back to Britain, employing tens of thousands of Iranians and keeping them in squalid conditions even as their British masters lived lavish lives. In 1917, Britain completed its conquest of Iran by consolidating its own interests with those of the Russians, who had held northern territory in the country but were struggling with their own revolution, abandoning their Persian claims to the British. In 1919, Britain enacted the harsh Anglo-Persian agreement, again by bribing Iranian negotiators, and took control of the country's army, treasury, transport, and communications systems. In 1935, the company changed its name to Anglo-Iranian Oil Company.

For the next three decades, Britain prospered on cheap Iranian oil while Iranians suffered its brutal occupation of the country. The British fleet now ran on oil, rather than coal, bringing Churchill's vision to fruition. However, discontent and nationalism rose in Iran after World War II, especially after the British government forced the brutal ruling monarch, Reza Shah, to abdicate in 1941, installing his son, Mohammed Reza Shah Pahlavi, as a puppet. In the ensuing decade, democracy began to take root, despite Britain's control, and in 1951, the Iranian parliament elected Mohammed Mossadegh, a strong voice for nationalizing Iran's oil resources, as prime minister. Within days, Mossadegh nationalized the oil reserves, ejecting the British and taking control of the oil. Incredulous, the British withdrew their technical people, froze Iranian bank accounts in Britain, and blockaded the ports, but Mossadegh ultimately prevailed. When it became clear that Mossadegh had succeeded in taking back control, the British government petitioned the United States to get involved, and, shortly after Dwight Eisenhower was elected president, the CIA was cleared to orchestrate a coup, dubbed Operation Ajax, to depose Mossadegh in mid-1953. During the turmoil, the shah fled to Europe, but he returned after the successful coup and was reinstated as monarch, immediately handing all control of Iran's oil assets back to the Anglo-Iranian Oil Company, which rebranded itself British Petroleum in 1954. The real tragedy was that not only did a US-backed coup depose a duly elected prime minister, it basically killed democracy in the country for generations.

Nationalism in Iran grew strident after the British retook control. The repressive regime of the shah caused great unrest in the country, and Islamic fundamentalism began its rise, finally exploding in 1979 with the Islamic Revolution, the final exile of the shah, and the takeover of the government by the Ayatollah Khomeini. The US embassy was stormed, the staff taken as hostages, and the extremely anti-Western Islamic regime nationalized its oil company again, this time ejecting the British from the country for good. Photos from the 1979 revolution showed crowds carrying huge banners with the likenesses of not only the ayatollah but also Mohammed Mossadegh, who had died under house arrest in 1967. The Anglo-Iranian Oil Company had already rebranded itself as British Petroleum; in 1998, after the merger with Amoco, it took on the name of BP Amoco for a short time, and finally, in 2000, it shortened the name to simply BP.[4-7]

Certainly, most of the unrest in the Middle East can be laid at the feet of the Americans and the British for their 19th- and 20th-century foreign policies. There aren't many, if any, companies who share that same blame, though, except for BP.

## The Troubled Rise of Modern-Day BP

In the following decades, BP was regarded as a second-tier international producer, prior to Lord John Browne's rise in the company in the 1990s. A career-long employee of BP, with the encouragement of his father, who hailed from the days of the Anglo-Persian Oil Company, Browne joined as an apprentice in 1966, rising steadily through the ranks and taking positions in business units in Europe and North America. By the mid-1990s, he was in the boardroom, becoming group chief executive of the combined companies after orchestrating the acquisition of Amoco in 1998. Under Browne, BP expanded its footprint, gobbling up competitors, and became well known for its brutal cost cutting. He was also one of the first greenwashers, pushing the public image that BP was a "green" company, unveiling the new daisy logo and the "Beyond Petroleum" slogan.

In 2001, BP "partnered" with the National Wildlife Federation to green up its image. For a donation, reported to be $113,000, the federation allowed its logo to be put on stuffed animals (representing the "Endangered Wildlife Friends") to be sold at BP-Amoco gas stations. The gas stations were plastered with posters of endangered animals (with appropriately placed BP daisy logos).[8] In 2002, BP launched a $200 million greenwash advertising campaign, complete with television ads and billboards in Times Square that shouted IF ONLY WE COULD HARNESS THE ENERGY OF NEW YORK CITY, and rolled out the now much-derided "Beyond Petroleum" slogan.

Lord Browne made public pronouncements about the dangers of global warming, touting the company's investments in solar, wind, and biofuel energy. In a case of unfortunate timing, at the same time that BP was touting their new image, they were also lobbying the US government to open the Arctic National Wildlife Refuge for drilling. Environmental groups charged hypocrisy, and protests in front of the St. James headquarters in London embarrassed Browne. The company quietly began pulling ads where they could, but the damage was done. Around the time Browne stopped publicly lobbying to open the refuge, Ronald Chappell, then head of press relations for BP in Alaska, asserted to the New York Times, "I think that the company has sort of decided that the role of corporations in public life is one of standing back and letting governments make decisions, trying to inform public policy but not making political contributions." That wasn't true, of course, with the company's employee PAC contributing $560,000 in 2000 and BP actively supporting Arctic Power, a pro-development lobby group that advocated drilling. BP quietly withdrew from that group, too, in November 2002.[9]

In the years since the company's rebranding, BP's investments in "green" energy have been paltry in comparison with their oil investment budget, even as the company has continued to grow traditional oil- and gas-based assets. Though they still use the "Beyond Petroleum" slogan, it rings hollow to an educated public that has witnessed environmental catastrophe and human trag-edy at the hands of the "green" energy company.

Browne's rapid global expansion, along with the cost cutting, is blamed for creating the conditions and culture that led to the 2005 Texas City Refinery explosion that killed 15. In 2006, the US Chemical Safety Board (CSB) issued a scathing report about BP's operation of the plant, which it had acquired in the Amoco merger, specifically blaming BP's management for ignoring major deficiencies in the plant. CSB chairman Carolyn W. Merritt said in a statement,

> The CSB's investigation shows that BP's global manage-ment was aware of problems with maintenance, spend-ing, and infrastructure well before March 2005. BP did respond with a variety of measures aimed at improving safety. However, the focus of many of these initiatives was on improving procedural compliance and reducing occupational injury rates, while catastrophic safety risks remained. Unsafe and antiquated equipment designs were left in place, and unacceptable deficiencies in preventative maintenance were tolerated.[10]

BP was ultimately fined $87 million for safety violations and paid over $2 billion in claims on over 1,000 lawsuits that were filed after the explosion.[11] This incident, combined with others, including an Alaskan pipeline spill blamed on reduced main-tenance to cut costs, began to tarnish Browne's "green" initia-tive. Combined with the deaths of BP employees, the accident compromised Browne's standing at BP, and it was announced he would retire in 2008. That was accelerated, though, to May 2007 after revelations in the British press about his secret gay lifestyle. Desperate to keep his private life out of the press, he ultimately lied in court about how he met one of his partners, and his career at BP came to an abrupt end. Tony Hayward, also a career BP employee, and a geologist, succeeded Browne as chief executive.[12]

## Could Different Leadership Have Spared the Gulf?

It's been said that since Browne had ruthlessly cleaned house of anyone who opposed him, Hayward was last man standing. According to BP critic and investigative journalist Tom Bower, "He's not very good."[13] Bower has criticized Browne for not promoting engineers to run the corporation, implying that if engineers had been running various operations, then the incidents like the Texas City explosion, multiple Alaskan pipeline leaks, and the Macondo blowout wouldn't have happened. I disagree. Bower's conclusion, one that's often drawn by industry outsiders, is based on limited knowledge of how oil companies are actually run. Certainly, engineering is key; but you don't have anything to engineer if geophysicists can't find oil-bearing subsurface structures to model, and if geologists can't model those structures for the engineers to target. You can't drill for those targets if the landmen haven't bought the leases, and you can't count the money the oil makes without accountants and finance types. The oil industry is very interdisciplinary. The best managers are those who immerse themselves in all aspects of the business, not just thump their chests and point to the diploma on their wall.

BP had certainly made poor management decisions over the years, but those weren't necessarily because engineers may or may not have been in charge. I've known many oil industry executives in my career. Certainly there are great CEOs who happen to be engineers; I also know some engineers who couldn't run a doughnut shop with a full-time management consultant and who have run their companies into the ground. Conversely, one of the best operations executives I've ever worked with was actually a degreed petroleum landman, and one of the best gas-gathering and gas-processing executives I know doesn't even have a degree. Ability to run an office, a division, or even a corporation has nothing to do with the diploma in that dusty frame hanging on your office wall; it has to do with your experiences, what you make of them, what's between your ears, and your leadership ability. Great

CEOs come from all disciplines. Saying that only engineers can run oil companies is like saying that only pilots can run airlines; we know how untrue that is.

BP's problem wasn't lack of engineers. Their real problem was a stifling culture built by a history of geographic domination, government-based bureaucracy, and, in Browne's day, an egocentric CEO who was a poor leader. When Hayward took the helm, he began working on BP's reputation and culture but was bogged down in decades of inertia and the cult of personality fostered by Browne. He was ill suited to the job of global energy CEO and utterly unprepared for the tsunami of oil in the Gulf of Mexico in which he was about to be swept up. Hayward's struggle and ultimate failure were emblematic of BP's own struggle to get control of this environmental catastrophe. The Anglo-Iranian Oil Company British Petroleum was unprepared to deal with political upheaval in Iran when they first lost control in 1953 and finally in 1979; they were again unprepared for the political and monetary upheaval caused by the Gulf blowout. It will forever change the face of BP; one can only hope that the change is for the better.

# | SIX |

# How Did We Get Here?
# A Brief History of Offshore Drilling

The common belief is that offshore drilling started only after easy onshore drilling opportunities were exhausted, but nothing could be further from the truth. Drilling in the water started very early on, when the Age of Oil was in its infancy. The first recorded well drilled in the water was actually around 1896 when Henry L. Williams, an oilman, as well as a spiritualist who held séances in the Santa Barbara town of Summerland, California, drilled the first reported offshore well from a pier he built 300 feet out from the beach. Earlier, as he drilled shallow deposits onshore, he noticed that the closer he drilled to the water, the better the wells were, so he first drilled on the beach and then moved to a pier. Others followed suit, and soon piers jutted out into the water as far as 1,200 feet, lined with derricks drilling for shallow oil sands. Over the next several years, 14 more piers were built out into the Santa Barbara Channel, holding more than 400 wells drilled in the shallow water. The field produced for 25 years, fouling the beaches and surrounding water.[1]

The next development in offshore drilling is still a bit disputed, even after more than 100 years. Generally, Caddo Lake in Louisiana is considered the first real stand-alone project drilled in water, starting around 1911. However, some make a remarkable claim that the first offshore well was actually drilled in Ohio, of all places. Back in the late 1830s, the then-largest man-made reservoir was dug in western Ohio, near the Indiana border and the towns of Celina and St. Mary's. The lake, about 7 feet deep, supported the construction of the Miami and Erie Canal that came through that area. It took 1,700 men nine years to excavate the lake, covering about 17,500 acres. About 40 years later, the Lima-Indiana oil trend was discovered, stretching 260 miles across Ohio and Indiana . . . with the new man-made lake lying right in the path. Small oil companies drilled right up to the shores of the lake and then finally out into the water, constructing their derricks on top of "cribs" built into the lakebed to make a foundation. At one time there were 100 wells in the shallow lake, but by the early 20th century, the boom was over as oilmen moved to Texas, Louisiana, and Appalachia.[2]

I'm not sure drilling in a 7-foot-deep artificial lake can be considered offshore, but it was done in the water, after all. Never the less, when Gulf Refining began drilling Caddo Lake in 1911, it was the first serious drilling in water, using tugboats and pile-driven platforms. After the first successful well, Gulf built a platform every 600 feet across the lakebed, the first time facilities were freestanding and not tied to shore by pier. During development, Gulf actually built a dam across the lake, once called Ferry Lake, to raise the level, making it easier to get boats to the locations in the water. In 1922, near what is now Baytown, Texas, oilman Jim Abercrombie began stepping off the shores of Galveston Bay in the Goose Creek field, drilling wells off wooden platforms built in the water.

## Moving into the Oil-Rich Gulf of Mexico

The first well in the Gulf of Mexico was drilled in 1938 from a freestanding facility built by Brown & Root (sound familiar?) about a mile offshore of Creole, Louisiana, in about 14 feet of water. A partnership between Pure Oil and Superior Oil, the Superior-Pure State No. 1 was a successful completion, but the platform was destroyed by a hurricane in 1940. Rebuilt and restored to production, the platform eventually produced 4 million barrels of oil from 10 wells drilled. Wooden platforms constructed as permanent drilling and production structures eventually gave way to the use of mobile drilling units that would be floated to a location, submerged to provide a solid drilling platform, then refloated to the next location. Drilling rigs and equipment were converted from land use to offshore, and gradually improved for drilling in a marine environment.

The year 1947 is considered the big kickoff for the offshore drilling industry. In that year, Kerr-McGee, Stanolind, and Phillips Petroleum partnered to drill the Kermac 16 well in Ship Shoal Block 32, 10 miles offshore in about 20 feet of water. Again, Brown & Root built the platform, which withstood one of the largest hurricanes of the season only a week after completion, surviving winds that were estimated at 140 miles per hour. This field was the first big discovery in the offshore Gulf of Mexico, producing 1.4 million barrels of oil and 307 million cubic feet of natural gas by 1984.

After the success of the Kermac 16, advances offshore came fast and furious. Even as Kermac 16 was being discovered, Superior was building another platform, this one 18 miles off Vermillion Parish, Louisiana. The Kermac well was the first step in that direction. The motors, fuels, and drilling supplies were kept on a converted navy barge moored at the platform, which held only the derrick and drawworks. It became the standard for offshore drilling in shallow water. One successful well followed another, and by 1949 the Gulf of Mexico had 11 operating fields with 44 exploratory wells.[3, 4]

As companies ventured farther out into deeper water, mobile drilling units began taking the place of the permanent platforms that were constructed to support the derricks. One of the most important developments in the offshore was the totally self-contained mobile drilling unit that could be moved easily from well to well. The first of those was a rig designed by A. J. "Doc" LaBorde, a naval engineer who had worked for Kerr-McGee after the war. While with Kerr-McGee, he designed a submersible barge rig that could drill in 40 feet of water, feed and sleep 58 men, and run round the clock on 12-hour tours rather than sending the crew in every night. Kerr-McGee rejected his design, saying it was too costly and impractical, so he left the company to build the rig on his own with new investors. In 1952, he pitched his idea to legendary oilman (and cotton ginner) Charles H. Murphy, who threw in with him and helped raise the rest of the money to build the rig. It was built at Alexander Shipyard in New Orleans, launching in 1953 and dubbed "Mr. Charlie" in Murphy's honor. The rig marked the beginning of another huge step into the offshore.

Shell Oil needed a rig to drill East Bay, just out of the mouth of the Mississippi River, and didn't believe it would be economical to drill using conventional permanent facilities. They contracted with LaBorde to drill the first well with Mr. Charlie. If it was successful, Shell would contract with LaBorde to drill the whole field; if they failed, it would be "Good-bye, Mr. Charlie." LaBorde later admitted that he had his own doubts of its capabilities, but he shouldn't have worried . . . Mr. Charlie was a smashing success and drilled not only the entire East Bay field but hundreds of other wells in the Gulf's shallow water before being retired in 1986. As rigs moved into deeper and deeper water, Mr. Charlie couldn't keep up and eventually became obsolete. Rather than send it to the scrap heap, though, like so many other rigs that lived beyond their productivity, industry leaders pitched in and gave it a permanent home at the International Petroleum Museum in Morgan City, Louisiana, where it lives today as the main attraction.[5]

Offshore development accelerated through the next few decades as more sophisticated mobile drilling units were developed. Jackup rigs were designed and built as the next technological step in mobile drilling unit technology. A jackup started out as a barge with legs that would be lowered onto the seafloor, some even using a mat the size of the entire barge for stability. Once on bottom, the engines on the rig would raise the barge on its legs above wave height and tides. Modern-day jackups use giant rack-and-pinion systems to raise themselves out of the water; the largest, called Gorillas and Super Gorillas, can drill in over 500 feet of water and drill below the seafloor to depths of below 30,000 feet.

As operators moved even farther out, submersible barges and

jackups didn't have the ability to keep up. In the 1950s, drillships were developed from old navy tenders, but in the days before deepwater mooring and stabilizing systems, roll and heave made drilling very difficult to control. In 1961, Blue Water Drilling Company and Shell Oil inadvertently invented the semi-submersible rig while moving the shallow-water submersible Blue Water No. 1 rig from one location to another. The rig was originally designed to sit on bottom after flooding its pontoons, so it could drill the well, and then be floated to the next location. With equipment on the deck, though, it couldn't float completely out of the water, so they moved it partially submerged. As they transported it, they noticed how stable it was with the pontoons below the water's surface, but still floating off bottom. They decided to drill with it this way, held in place by anchors, and the semi-submersible was born.[6]

Within 10 years, 30 purpose-built semis had been constructed. They now dominate the deep- and ultra-deepwater, prized for their stability in rough seas. Drillships were also greatly improved during these years, often used for drilling initial exploration wells in ultra-deepwater. Transocean's *Discoverer Enterprise*, which was on station at the Macondo site for months as BP struggled to get control of the blown-out well, is a prime example of this type of drilling unit. Built in 1999, the *Enterprise* is 835 feet long, carries a double derrick—which means it can work on two wells at the same time—and can drill in 10,000 feet of water. It also has the ability to process around 25,000 barrels of oil per day. Like the *Deepwater Horizon*, it uses dynamic positioning, so it doesn't require anchors or a mooring system to stay on station.[7]

## An Ever-Expanding Industry Leads to
## New Laws, New Technology, and New Debates

As oil companies pushed farther and farther from shore, the US government hustled to keep up. Several important pieces of legislation were passed at around the time Mr. Charlie went into service, starting with the US Submerged Lands Act, passed in 1953, which established the 3-mile limit delineating state lands from federal lands. In the same year, the Outer Continental Shelf Lands Act was passed, giving the Department of the Interior authority over managing these lands for the US government and collecting royalties for oil and gas production from the outer continental shelf. The first offshore lease sale to private companies was also held that year. After the Santa Barbara spill in 1969, Congress passed the National Environmental Policy Act, which required detailed government review of projects for environmental impact, along with the Clean Air Act, which regulates air emissions from rigs, gas plants, pipeline stations, and other industrial installations. In 1972, the Coastal Zone Management Act, administered by the National Oceanic and Atmospheric Administration (NOAA), was passed; this requires both state and federal consent for actions that would affect land and water use of a defined coastal area. In 1977, Congress passed the Clean Water Act, which regulated the discharge of pollutants into surface waters and established fines for violations.[8]

Offshore activity continued unabated throughout this period. In 1982, Congress passed the Federal Oil and Gas Royalty Management Act, which mandates protection of the environment and conservation of federal lands in the course of building oil and gas facilities. The secretary of the interior was given authority to establish the Minerals Management Service as the administrative agency responsible for the mineral leasing of submerged outer continental shelf lands and for the supervision of offshore operations after lease issuance to operators.[9] In 1983, after the Law of the Sea Conference held the previous year, President Reagan issued a proclamation declaring the US Exclusive Economic Zone extending out 200 nautical miles from our coastlines, specifically claiming for the United States exclusive use of mineral resources that go beyond the outer continental shelf.[10] The legal groundwork had been established for the development of deepwater oil and gas resources.

In 1989, the *Exxon Valdez* disaster in Alaska's Prince William Sound—the largest oil spill in history until the BP disaster—drove the government to legislate more safety, environmental, and regulatory oversight. In 1990, the Oil Pollution Act established the federal framework for preparing for oil spills, establishing responsibilities, fines, and readiness requirements for companies operating in the offshore. It established the national Oil Spill Liability Trust Fund and established a three-tiered approach to oil spill management composed of federal, state, and local government as well as operators of offshore facilities or vessels that could cause an oil spill.[11] Responding to growing environmental concerns about offshore drilling, President George H. W. Bush placed a 10-year blanket moratorium on new-area offshore leasing.[12] In 1994, Congress closed the Outer Banks of North Carolina, and California closed its state waters to new drilling.[13]

## High Technology Takes Over

In the meantime, during the 1990s, drilling in the western Gulf of Mexico increased as 3-D seismic gave new life to finding "bright spot" anomalies, given that name by geologists who searched for the characteristic bright colors that hydrocarbon-filled subsurface structures exhibit on seismic interpretation maps. As these plays, as they are called, were drilled up in shallow water, new territory opportunities moved into ever-deeper water.[14]

As operators moved farther out, new technology was developed for subsea well control; blowout preventers, wellheads, and production units were designed to remain on the seafloor, controlled remotely from the surface. A whole new industry grew up around deepwater drilling, with companies like Cameron, Hydril, Halliburton, Weatherford, and Oceaneering, as well as others, leading the way in high-tech remotely operated subsea well control, intervention, production, gathering, and processing. Drilling technology also grew by leaps and bounds with the advent of synthetic drilling fluids, LWD (logging while drilling), and sophisticated drilling methods guiding the drill bit to very precise targets miles away from the rig's surface location. Massive floating production facilities were built, capable of processing tens of thousands of barrels per day from multiple wells and delivering that production into a huge offshore infrastructure that brought it onshore for delivery to end-user markets.

Once exploration companies stepped off the outer continental shelf, they first probed the flex trend—the slope between the margins of the shelf and the deepwater itself—and then ultimately the deepwater, defined as depths greater than 1,000 feet, and going as deep as almost 10,000 feet. Operators routinely drill in 5,000 feet of water and deeper as they expand their exploration plans. Since 2002, 6.6 billion barrels of oil have been discovered in the deepwater.[15] The deepwater is productive from Texas to Alabama, and these regions have recently dominated offshore US production as new-build floating rigs have given operators more opportunities to explore the deeper reaches of the Gulf of Mexico. Tiber, the prospect drilled by the *Deepwater Horizon* before moving on to the Macondo well in January 2010, is located 250 miles southeast of Houston, Texas, in over 4,000 feet of water, and was drilled to over 35,000 feet.[16] Deepwater prospects extend from that region, called Keathley Canyon, eastward across the Gulf, accounting for the vast majority of new drilling in US waters.

In 2008, responding to skyrocketing gasoline prices, President George W. Bush saw an opportunity to reverse his father's executive ban on new-territory offshore leasing and lifted it on July 14, 2008.[17] Well, actually, he didn't lift it. In a statement he challenged Congress by saying essentially, "If you go first and lift the ban, so will I." In September of that same year Congress did just that, allowing the ban to expire quietly during a contentious election season when voters were suffering with record-high gasoline prices. It wasn't until March 31, 2010, though, that executive action again moved to affect the range of offshore drilling. That's when President Obama announced his intention to expand offshore leasing to the eastern Gulf of Mexico, beyond 125 miles off Florida, the southern Atlantic coast, and the Alaskan coast. He was rewarded for that step with the blowout of the Macondo well three weeks later, likely scuttling offshore expansion for another generation.[18]

Today about one-third of our domestic oil production comes from the outer continental shelf,[19] and about 80 percent of that comes from the deepwater.[20] Royalties from oil and gas produced on federal lands have become a key source of government revenue. Clearly, the deepwater Gulf of Mexico has become a key source of our energy supply, even as our dependence on fossil fuels continues to grow.

BP's Macondo well blowout has cast a pall over the deepwater, raising questions about the safety and sustainability of this source of energy supply. As we continue to be dependent on fossil fuels as a prime source of energy, our energy security hangs in the balance. A permanent shutdown in the deepwater would cause us to increase imports in coming years as production from these wells, which rapidly declines, begins to tail off with no new production coming on. A side effect (no small one, I might add) would be the damage to the Gulf Coast states, as thousands of jobs and hundreds of millions of dollars in corporate revenues would be threatened.

In its present-day form, the offshore oil and gas industry represents the collision of ingenuity, technology, profit motive, and poor government policy. As generations of elected leaders failed in their responsibilities to plan for our energy needs, private industry has stepped into the breach for profit. We, as a nation, now burn more oil per person than any country in the world, and we do little or nothing to curtail it. The US government has become dependent on oil royalties to fund its mammoth expansion and war fighting, and Americans have come to look at their ability to fuel their SUVs with cheap gasoline as a right that is as fundamental as free speech. The fruits of such irresponsible, decades-long buck passing are catastrophes like the *Exxon Valdez* and now the *Deepwater Horizon*. If we don't change the path that we are on and learn from our own history, we are simply doomed to repeat it.

# SEVEN

# Corporate Shape-Shifting
# and the BP–Government Merger

One of the most difficult things to figure out during the early weeks and months of the spill was who was really calling the shots on the response—BP or the government. It was clear that the government didn't have the expertise to manage the blowout on the seafloor and that the beaches were being guarded by both private and local government officials who answered to BP. Though the president and other federal government agency heads continually told the media that the government was in charge, that fact wasn't readily apparent on the ground. As the crisis wore on, the relationship between BP and the US government became closer as the individuals involved came to know one another, and they collaborated to get the well under control and somehow contain the massive environmental damage.

It's hard to pinpoint when the merger between BP and the US government in their response to this catastrophe occurred, but it became obvious after BP agreed to commit to a $20 billion fund for spill cleanup on June 16, 2010.[1] On federal lands, the relationship between the US government, as regulator, and the industry has always been one of familiarity. In the small towns located in producing areas, like the Gulf Coast, often the regulator lived next door to the regulated, and a neighborliness developed because of the close proximity. It is difficult to come down on a company represented by the guy who coaches your kid's Little League team, so the government became somewhat of a benign overseer. Agency management often referred to the companies being regulated as "partners" or even "customers," ceding the authority necessary to truly be the watchdog they were supposed to be. After the blow-out, the government wanted to be seen as the one in authority. The problem was, though, that it lacked the skills or resources to actually do anything and was completely dependent on BP for the technical expertise to manage the well. A co-dependent relationship was formed that went beyond traditional industry–government coziness. This relationship between the government and BP morphed into the odd partnership out of necessity that was beneficial to both entities. To understand this odd partnership, you have to go back to the early hours of the disaster, when lines of authority were originally established during the direct response to the blowout, the fire, and the missing people.

## The Early Hours

As the *Damon Bankston* was taking on survivors from the fast rescue craft and the rig's lifeboats, the Coast Guard arrived by helicopter. The rig crew head count was short, and—not knowing who was where—responders began their search for survivors immediately as they counted noses again and again. The Coast Guard began ferrying the injured to hospitals on shore, and the search for those missing widened—until the 11 left unaccounted for were presumed dead due to their last known locations on the rig. The fire, in the meantime, raged on.[2]

Though the Coast Guard was on location, it did not take charge of the scene or lead the firefighting effort—neither of which is part of its primary mission, especially since the service was moved into the Department of Homeland Security (DHS) after 9/11. Following the attacks on that tragic day, the Coast Guard's emphasis on search-and-rescue and marine safety was reduced and its law enforcement responsibilities increased. Those changes further moved the Coast Guard away from its historic role—one that had already been de-emphasized when its drug interdiction duties were increased after the War on Drugs was declared by Richard Nixon in 1972.[3] Under authority of the DHS, the Coast Guard is now considered the last line of defense in our ports against terrorism and smuggling. It also maintains its traditional marine-safety presence. Budget constraints, as well as its widening mission, have stretched the Coast Guard to the point that Admiral Thad Allen, who was commandant prior to retiring in May, had declared earlier that the new responsibilities threatened to make the service a "hollow force" if its fleet of aging ships wasn't upgraded. The current average age of the Coast Guard's large cutters is about 41 years, compared with the Navy's 14. Allen warned that the age of the fleet is "putting our crews at risk and jeopardizing the ability to do our job."[4]

Even so, the early response of the Coast Guard, like the performance of the crew on the *Bankston,* was exemplary. But weaknesses in offshore rig safety responsibilities and regulations became readily apparent. Currently, owners and operators of offshore rigs are solely responsible for primary firefighting, and, under brand-new rules, the Coast Guard is to not get involved beyond supporting an expert fire marshal brought in by the company whose rig is on fire. In the confusion of the night of the blowout, no one called a fire marshal in, and the vessels surrounding the rig started pouring tons of seawater on its decks with onboard water cannons. Daun Winslow, a Transocean executive who had been visiting the rig when the blowout occurred, was concerned about the *Horizon*'s stability and gave early instructions by radio to the firefighting vessels to put water only on the columns that supported the rig to keep them as cool as possible, and to avoid flooding the decks.[5] However, because there was no one on the scene who was really in charge, including the Coast Guard, that request was ignored. The private boats poured tons of water directly onto the decks, flooding the vessel and upsetting its stability. Over the next 36 hours, the rig listed farther and farther as the boats continued to flood it. As we all know, the *Deepwater Horizon* sank on the afternoon of the 22nd, taking with it the riser that had connected the rig to the well.

Had responders been able to keep the rig afloat longer, most of the oil would have been burned at the surface, possibly giving Transocean more time to get the riser separated from the blow-out preventer on the bottom. Poor communication and coordination between Transocean and its contractors, as well as the Coast Guard's hands-off approach, certainly accelerated the loss of the rig.[6]

During this time, BP was nonexistent on the scene, as it was scrambling to coordinate its own response and trying to distance itself from the blowout and Transocean. It didn't take long for the finger pointing to begin, though. In a press release on April 21, the company, referring to the blowout as "a rig fire," said it "stood ready to assist in any way in responding to the incident." Then-CEO Tony Hayward, from BP's St. James offices in London, said in a statement, "Our concern and thoughts are with the rig personnel and their families. We are also very focused on providing every possible assistance in the effort to deal with the consequences of the incident."[7] (Of course, only the London press office contact was provided in case someone actually tried to call for assistance.)

This is the characteristic response of BP, the old arm's-length strategy, but it didn't work for long this time. Senior management at BP certainly knew that this wasn't just a "rig fire," unless London was far more out of touch with their American subjects than we all thought. This was a catastrophic blowout that had cost 11 lives, and they already knew it. Punting the ball to Transocean was a rapidly formulated tactic that worked for only a couple of days, until the facts surrounding the blowout started leaking out from witnesses and other sources.

### The Early Days

The second day of the blowout, April 22, ironically happened to be the 40th anniversary of Earth Day. In a perfect example of *your-timing-couldn't-be-worse-even-if-you-tried*, about an hour after the burning rig slipped below the surface of the Gulf, the president gave a tone-deaf Rose Garden address that completely ignored the unfolding crisis. During the speech, he recognized guest Dennis Hayes, who as a graduate student organized the first Earth Day for Democratic senator Gaylord Nelson in 1970, and then spoke of our future in clean energy, saying,

> With your help, we've made a historic investment in clean energy that will not only create the jobs of tomorrow, but will also lay the foundation for long-term economic growth. We've continued to invest in innovators and entrepreneurs who want to unleash the next wave of clean energy. We've strengthened our investment in our most precious resources—the air we breathe, the water we drink, and the parks and public spaces that we enjoy.

He went on to say,

> I think we all understand that the task ahead is daunting; that the work ahead will not be easy and it's not going to happen overnight. It's going to take your leadership. It's going to take all of your ideas. And it will take all of us coming together in the spirit of Earth Day—not only on Earth Day but every day—to make the dream of a clean energy economy and a clean world a reality.[8]

As the president spoke about clean energy and a clean world, the *Deepwater Horizon* was settling to the bottom of the Gulf of Mexico, having just separated from a roaring subsea well that was about to cause the worst environmental catastrophe in US history. Eleven men were dead. The well was flowing uncontrolled 5,000 feet below the surface of the water. And the president didn't even mention it.

Later that day, White House Press Secretary Robert Gibbs issued a benign statement that the president had been briefed that day by DHS Secretary Janet Napolitano, Interior Secretary Ken Salazar, Admiral Thad Allen, who was still Coast Guard commandant at that time, EPA Deputy Administrator Bob Perciasepe, and FEMA Administrator Craig Fugate, as well as White House staffers. The lightbulb had just come on that this was a serious incident in the making, but the reality of the situation had not sunk in even yet. Salazar stayed in Washington and dispatched a deputy to the area to "coordinate" efforts.[9] Unfortunately, since the Obama team has few close advisers who are businesspeople and is openly contemptuous of anyone from the oil and gas industry, there was no one in the room, or indeed in Washington, who understood what was getting ready to happen.

The Coast Guard, beginning to realize the possible magnitude of this catastrophe, finally swung into action with a more robust response after the rig sank. Using Shell's donated office in Robert, Louisiana, they set up a command center and established the Unified Command—a cooperative organization designed to coordinate the response to the incident. It consisted of multiple federal agencies, as well as the Coast Guard, BP, and Transocean. In the early briefings, BP and the Coast Guard's Admiral Mary Landry stood shoulder-to-shoulder as they addressed a news-hungry press corps who didn't know what questions to ask, most of whom would print whatever they were told from the podium. One of the first announcements made by Admiral Landry was that oil flow had stopped, which both BP and Transocean already knew was completely false.[10] Even as she was making the statement, BP had remotely operated vehicles on the seafloor, desperately trying to shut in the well through the intervention panel on the BOP using what are called hot stabs, or devices that the ROV inserts into ports on the panel to operate the rams. Oil from the well was roaring out of cracks in the bent-over riser less than 30 feet above the ROVs, so BP knew the well was not shut in, but that didn't stop them from standing mute while the Coast Guard gave the American public falsely comforting information.

BP engineers continued to struggle for a day to shut the well in with ROVs, only to discover that they could not. When Transocean brought in drawings of the BOP that had been built by Cameron International 10 years earlier, they didn't match what the ROVs were seeing on the well. It turned out, of course, that the Macondo's BOP had been extensively modified. After unsuccessfully trying to close the middle pipe rams on the drill pipe that was still in the BOP, they discovered that the intervention panel had been replumbed; they were actually firing the test ram at the bottom of the stack, which could not hold pressure or close around pipe. The middle pipe rams wouldn't close. As they struggled for days after the blowout with the BOP, which was years overdue for its inspection and recertification, BP and Transocean said nothing to the public.[11]

On April 25, Admiral Landry reluctantly admitted that the well was indeed flowing, but she announced a ridiculously low flow rate of 1,000 barrels per day—an estimate that no reasonable person would have believed.[12] The struggle with the failed BOP continued, and nothing worked. BP representatives continued to stand silently during the briefings, allowing the Coast Guard to continue giving false and misleading information to the press. During this same time, BP finally began showing 10- to 15-second video snippets of ROV activities on the seafloor (and the smallest of the three riser leaks), giving no meaningful information and refusing to talk about how much oil was pouring from the well. The pattern of misinformation/no information from BP was established. After almost 100 years, the DNA of the Anglo-Persian Oil Company still informed its descendants' behavior.

As the oil slick on the surface continued to spread, it became obvious: The well was flowing far more than 1,000 barrels per day. While BP remained publicly silent on the issue, the Coast Guard reluctantly raised its estimate of flow to 5,000 barrels per day on April 28.[13] No one but a few in the media believed that number, either. Those of us in the industry understood that the flow rate was still multiples more than even this estimate, knowing that strong deepwater wells often come in, under *controlled* conditions, well over 20,000 barrels per day. Uncontrolled, this well would be flowing more than that . . . a lot more.

## Reality Sets In

In the early days after the blowout, the federal response was slow. Even though the Coast Guard was involved from the first hour, and the president was being briefed regularly, the administration quickly fell behind the crisis, embroiled in the normal day-to-day in Washington. Then the Gulf states' governors started calling. The media started criticizing. Congressman Ed Markey of Massachusetts, chairman of the House Energy Independence and Global Warming Committee, seemed to be the only one really paying attention at the time. No friend to the oil industry, Markey began making public statements about the oil spill and BP's handling of it. On April 29, he called the CEOs of Exxon Mobil, Shell, BP, Chevron, and ConocoPhillips for a public hearing about the spill and recent profitability of the industry.[14] That same day, Napolitano declared the BP disaster a "spill of national significance,"[15] and Bobby Jindal, governor of Louisiana, declared a state emergency.[16] The next day, during an interview on ABC's *Good Morning America,* the president's senior adviser, David Axelrod, announced that the Obama administration was shutting down all new-area offshore drilling—kicking off both concern and confusion in the industry.[17] Two days later, Napolitano named Admiral Allen as national incident commander, in charge of the federal government response. With Allen's appointment, the government began to take an active role in managing the disaster. A few days later, Markey began loudly talking about the use of dispersants, offshore drilling, and BP's spill response. After a private briefing with BP, Transocean, and Halliburton on May 4, Markey was very frustrated with the lack of answers he was getting and said,

> We asked a lot of questions; we did not get all of the answers that the American people, but especially the people in the Gulf region, really deserve. BP is now known as British Petroleum. If this leak is not shut off soon, it will become known as "bayou polluter," and they will be known that way forever.[18]

Even as the White House began to step up its involvement, defining BP as the responsible party, the president and his spokespersons continued to refer to what BP was doing, and would be doing, and defining the government role as essentially monitoring the activity—in other words, acting as third-party watchdog rather than taking authority. On May 2, during his first trip to the Gulf Coast, Obama reiterated that BP would pay for this entire cleanup, even though, by law, their liability was limited to $75 million. The next day, BP reaffirmed that it would pay more than its legal limit but hedged a bit in a statement, saying,

> BP is committed to pay legitimate and objectively verifiable claims for other loss and damage caused by the spill—this may include claims for assessment, mitigation and clean up of spilled oil, real and property damage caused by the oil, personal injury caused by the spill, commercial losses including loss of earnings/profit and other losses as contemplated by applicable laws and regulations.[19]

Tony Hayward, who would later be ousted as CEO over his handling of the spill, began a stumbling public relations campaign that probably raised more questions than it answered, repeating a phrase that was to become a mantra, "We will pay all legitimate claims." What did "legitimate" mean? BP was clearly leaving a door open. The more Hayward and other BP executives said it, the less the public believed it, especially as oil began coming ashore in Louisiana. Hayward became a one-man gaffe machine over the following weeks, with a few beauties like "the Gulf of Mexico is a very big ocean," and "the amount of volume of oil and dispersant we are putting into is tiny in relation to the total water volume." He also claimed that the impact of the massive oil spill would be "very, very modest."

The doozy that did him in, though, was an offhand comment he made to Louisiana reporters on May 30; in talking about how long it was taking to shut in the well, he blurted out, "There's no one who wants this over more than I do. I would love my life back." That did it; his unthinking comment about getting his own life back, when 11 people had lost their lives and millions had been severely impacted by the spill, was his ultimate undoing. Shortly after, he was shipped back to the UK, where, in another public relations gaffe, he was photographed riding in a yacht race that next weekend. In a damage control move by the company, he was ultimately replaced by American Bob Dudley.[20]

By early May, the response effort in BP's command center in Houston was frantic. They began designing and trying everything they could think of, and, as each remedy failed, confidence that they knew what the hell they were doing faded. The spectacular failure of the containment dome, then the marginally effective riser insertion tool, fed that sinking feeling that the public, the government, and even BP was beginning to feel. The *New York Times* later quoted Ken Salazar as saying, "There was an arc of loss of confidence. I was not comfortable they knew what they were doing."[21]

## The Finger Pointing Begins

On May 11, public testimony of industry executives before Congress began, kicked off by the Senate Energy Committee. Lamar McKay, chairman of BP America, Tim Probert, president of global business lines for Halliburton, and Steven Newman, CEO of Transocean, all pointed fingers at one another in the first hearing, but it was clear that BP was on the hook.[22] McKay, when pressed by senators for details about expenses BP would pay, kept repeating the mantra, "BP will pay all legitimate claims, BP will pay all legitimate claims." Key information did begin to surface during testimony, though, as the executives admitted "anomalous" pressure readings during the negative test, which was the first public acknowledgment by any of the companies that a mistake had been made while preparing to temporarily abandon the well.[23] The next day, the oversight subcommittee of the House Energy and Commerce Committee paraded the executives before the public again, and Chairman Bart Stupak ·of Michigan pointed out in his statement that his investigation revealed there were four major flaws in the blowout preventer that had yet to be acknowledged by Transocean: a hydraulic leak, extensive modifications, the fact that BOP can't cut all pipe (which we already knew), and the failure of the emergency controls.[24] We now know from later testimony that all the assertions in Stupak's statement were correct. As a matter of fact, we also know from later testimony that BP and Transocean knew about the faults in the BOP as early as April 21 and 22 but withheld the information from the public.[25]

On the same day as the Senate hearings, a joint investigation between the Coast Guard and the MMS began in Kenner, Louisiana, to probe into the causes of the blowout and loss of the rig.[26] While the congressional committees were mostly for show and sport-shaming the CEOs, the joint investigation was a serious technical probe of actual causes. These hearings proved to be most enlightening, as key players testified over the following months, under sometimes very difficult questioning by the committee and by attorneys for the companies involved, the flag nation of the Marshall Islands—where the *Deepwater Horizon* was registered—and persons of interest (those who were directly involved in the blowout or decisions that led up to the incident). At this writing, Coast Guard Captain Hung Nguyen and David Dykes of the MMS (now BOEMRE) have done a very good job of managing this complex investigation. However, they almost lost control of it in July as more and more attorneys for persons of interest became involved in these later sessions; a retired federal judge, Wayne Anderson, was added in August to act as referee. He got the hearings back on track by controlling (sort of) the antics of all the lawyers involved.

## Obfuscating the Volume

On May 14, BP finally got their riser insertion tool, or RIT, into place. The company claimed that the RIT—a siphoning pipe extending from a drillship and connecting to the broken, oil-gushing riser on the ocean floor—could collect up to 5,000 barrels of oil per day. And indeed, after several days of ramping up the RIT's production, BP spokesman Mark Salt announced on May 20 that they were "now capturing 5,000 barrels per day of oil."[27] At this point, the public remained unaware—though there's evidence that BP did not—that far more than 5,000 barrels a day was flowing from the blown-out well.

Yet BP continued to downplay the well's output—despite mounting evidence to the contrary. In a conference call the day after Salt released the 5,000-barrel-a-day number, BP COO for Exploration and Production Doug Suttles denied Salt's statement and claimed that the flow rate was actually just over 2,000 barrels per day. As the story changed, Suttles said (in what was becoming to onlookers his familiar you're-too-stupid-to-understand tone), "We never said it produced 5,000 barrels a day. I am sorry if you heard it that way."[28]

Actually, I think we—the media and analysts closely following the story—heard it that way because that is exactly what Salt said. But by this time, BP's shape-shifting was well established. To be fair, some of the discrepancies in disclosure were between actual oil gathered versus the flow rate at any one time, but the bottom line was that BP was containing only a small portion of the flow once they got the RIT in place and operating—and they weren't disclosing that reality to the public.[29]

That fact was painfully clear, though, to observers who were studying the video feed of the RIT in action—a feed released a couple of days before Salt's statement. That feed showed oil roaring out around the RIT. Oil was also flowing out of cracks in the wrecked riser just above the blow-out preventer. This alone was reasonable evidence that the well was producing much more than the RIT could capture—in other words, much more than the 5,000 barrels a day that remained the official number.[30] Yet BP continued to downplay the well's output.

As the weeks dragged on, though, even BP officials started cozying up to the 5,000-barrel-a-day number, likely because they knew the real output was much higher. They probably had calculated the actual flow rate to within a few percent, but they weren't about to tell anybody. In fact, they continued to assert that they weren't calculating the volume; even so, they would also argue every time a larger number was put out there, like the 70,000 barrels per day estimated by Purdue University professor of mechanical engineering Steve Wereley, who calculated the flow by studying video of the leaking riser.[31] On May 14, during an MSNBC interview, Robert Dudley—then BP's managing director and now CEO—took issue with the 70,000-barrel-per-day estimate, saying that it "feels like a little exaggeration, a little bit of scare-mongering." After again saying it was impossible to gauge the amount of oil, he maintained that "five thousand [barrels] is a good estimate."[32]

It was this attitude of BP officials that was so infuriating. Everyone who was paying attention knew that BP had two goals: Get the well shut in; and get the well shut in before the company was forced to measure the actual flow. Why? Liability is based on the amount of oil released into the environment. If you kept those two goals in mind when observing all of BP's actions and listening to their statements, you would understand why they did what they were doing, even when it seemed illogical at the time.

Perhaps in time it will also prove to have been illegal.

Throughout this entire debacle, all of BP's officials would continue to claim that calculating flow rate was impossible, that the company's response would have been the same no matter what the rate, and that knowing the flow rate was just not important anyway. However, two letters that surfaced in August through our old friend Ed Markey paint a different picture of what the company actually knew at the time they were claiming ignorance of flow rate. These letters, from Doug Suttles to Coast Guard Rear Admiral James Watson, dated July 6 and July 11, sought approval to apply dispersants at the wellhead using an oil flow rate to calculate the dispersant amount. The flow rate that Suttles used for the calculation? 53,000 barrels per day.[33, 34] They knew all along, but kept it to themselves.

## Obama Gets Tough (Finally)

During the middle of May, as oil roared into the Gulf and everything BP was trying was not working very well, the pressure on the Obama administration mounted. The president, not one to allow himself to get emotional, began to sharpen his rhetoric about BP. During a May 14 statement, Obama said,

> It would also help ensure that companies like BP that are responsible for oil spills are the ones that pay for the harm caused by these oil spills—not the taxpayers. This is in addition to the low-interest loans that we've made available to small businesses that are suffering financial losses from the spill. Let me also say, by the way, a word here about BP and the other companies involved in this mess. I know BP has committed to pay for the response effort, and we will hold them to their obligation. I have to say, though, I did not appreciate what I considered to be a ridiculous spectacle during the congressional hearings into this matter. You had executives of BP and

Transocean and Halliburton falling over each other to point the finger of blame at somebody else.[35]

The tone sharpened even more after BP agreed to put up a live feed of the blowing-out well putting thousands of barrels of oil into the Gulf. The pressure was mounting for somebody to do something, and, on May 27, after the top kill procedure had begun, Obama lashed out at the company,

> As far as I'm concerned, BP is responsible for this horrific disaster, and we will hold them fully accountable on behalf of the United States as well as the people and communities victimized by this tragedy. We will demand that they pay every dime they owe for the damage they've done and the painful losses that they've caused. And we will continue to take full advantage of the unique technology and expertise they have to help stop this leak.

Then,

> BP is operating at our direction. Every key decision and action they take must be approved by us in advance. I've designated Admiral Thad Allen—who has nearly four decades of experience responding to such disasters—as the National Incident Commander, and if he orders BP to do something to respond to this disaster, they are legally bound to do it.[36]

The battle lines were being drawn. BP supporters began to cry foul, complaining that the president's rhetoric was affecting the company's stock price. The mayor of London, Boris Johnson, even said that Obama's words were "anti-British rhetoric."[37] Investors and media began worrying about BP's viability and whether the company would even survive this fiasco. BP's stock plunged as the war of words continued, especially after the much-ballyhooed and much-delayed top kill failed a few days later, halted by Secretary of Energy Steven Chu.[38]

In the days after the failed attempt, the rhetoric pouring out of Washington against BP and the oil industry was withering. On May 30, Ed Markey turned up the heat to white hot during an interview on CBS when asked about BP's statements:

> I think they were either lying or incompetent. But either way, the consequences for the Gulf of Mexico are catastrophic.

And:

> I have no confidence whatsoever in BP . . . I do not think they know what they are doing. I do not think people should really believe anything BP is saying in terms of the likelihood of anything that they are doing is going to turn out as they predicted.[39]

Talk in the business media turned to Chapter 11 bankruptcy protection or BP being swallowed up by another Supermajor like Exxon Mobil or Shell. In the *New York Times*, Andrew Ross Sorkin reported that investment bank Credit Suisse estimated the liabilities from the spill at $23 billion for the cleanup. On top of that could be another $14 billion in claims from damaged businesses and fishermen; $40 billion in costs was not considered beyond imagination.[40] Something needed to be done, and fast.

## The BP–US Government Merger

On June 16, President Obama, Tony Hayward, and BP Chairman Carl-Henric Svanberg struck a deal: BP was to commit $20 billion over four years to cover the cleanup and damages. The fund would be independently managed by Ken Feinberg of 9/11 fund fame, and BP agreed not to cap the liability; all environmental fines would be paid separately. The company also agreed to fund a $100 million fund to help offshore workers who might lose their jobs during the moratorium. To help cover the costs, BP announced it would pay no dividends for 2010. What was not publicly disclosed was that President Obama and the federal government also agreed to get off BP's ass, which now seems obvious. The rhetoric immediately cooled, BP faded into the background, and Admiral Allen became the spokesman for everybody involved, save the occasional technical briefing from BP.

One of the reasons for the deal was obviously to assure that costs were covered. To do that, Obama had to keep BP in business in the United States. The other reason that the company was motivated to enter the agreement was that there had been rumblings from Congress about denying BP further drilling permits or barring them from operating in the US. In BP's announcement about the deal with the Obama administration, the company emphasized that one of its terms was that it would back the $20 billion with US assets.[41] In other words, the deal guarantees that BP will remain viable in the US since they need assets to pay the costs. It was actually a win–win arrangement. The only blemish on the White House love fest that day was the gaffe committed by BP Chairman Svanberg, a Swede, when he publicly expressed his concern for the "small people" on the Gulf Coast.[42] Most people gave him a pass, though; it's amazing how $20 billion makes folks a little more forgiving than usual.

The $20 billion commitment calmed the stormy seas (pardon the pun) of this disaster. Though not intended to cover all the environmental fines, the commitment silenced the critics who were forecasting apocalyptic consequences for the company and the damages it had caused. I believe that, had BP not stepped up to this commitment, the company could have been easily devoured by a larger predator, like Exxon, which would not have felt the same urgency to cover the damages. During the *Exxon Valdez* debacle, Exxon proved a formidable opponent, willing to wait as long as possible before paying damages, hoping that victims of that spill would die before the dough had to be ponied up. I would expect no different from a successor to this catastrophe.

In the months following, save for a few gaps in communication, BP and the Unified Command walked in lockstep. When BP stepped out of line, Admiral Allen was quick to assert his authority, but much of what he said to the public was obviously fed to him by BP. As the weeks dragged on after the well integrity test and relief well were delayed by weather, more tests, and indecision, Admiral Allen began giving almost incomprehensible press briefings as he tried to explain how everything was okey-dokey, but they were going to do more "static tests" anyway. Here's an example of this kind of doublespeak from his August 11 presser:

> Sure, there's a very low probability that we might have actually sealed the annulus with the cement that came down the pipe casing and came back up around it. What we want to do is understand whether or not there's what we call free communication. In other words whether there, the hydrocarbons in the reservoir can actually come up through the annulus outside the casing, if that's the case when we go in and we drill in we put the mud and cement we're just going to drive that down and seal the well. OK? If there's cement there and there's no communication that means we have what we call stagnate oil trapped around that casing up to the well head. If you go in and you start pumping mud and cement in there the chances are you could raise the pressure and push that up into the blow out preventer. And that's a very low possibility, low probability event but we want to, we want to test the pressure in the blow out preventer and see if we actually have pressure coming up that would indicate that we have free communication with the reservoir. If not that would change our tactics and how we do the final kill.[43] ·

It was obvious that BP and the science team had no idea where the cement during the "static kill" actually went. The explanation from Allen made no sense to anyone who knew what to listen for, and it was clear to me that he was repeating talking points rather than actually understanding what was going on. Since both the production casing and the annulus were exposed to the BOP, they didn't really know where they were pumping the mud and cement, save for some estimated pressures, but that didn't stop the admiral from making public statements supporting the contention that BP and the government "science team" actually did. My question was, though, if they were so certain, why did they do "ambient pressure" tests for days on end?

While watching all these machinations, it's been difficult to judge the relationship between Steven Chu, his "science team," and BP's technical folks from the outside, but a few clues have given some insight into the ongoing relationship between the two entities. One thing is clear: BP has been more than happy to take the public backseat while Chu and his team have been seen to be calling the shots. As an example, Joel Achenbach of the *Washington Post* reported that it was actually Chu's idea for the "well integrity test" when BP finally got the capping stack set in mid-July. The weeks of delays in getting the relief well completed were also a result of this odd partnership that was formed out of necessity. Admiral Allen, when asked about whose idea was for a certain procedure, answered, "It's hard to say anymore, we've been together so long. Some of these conversations start around the coffeepot."[44]

As the end of the summer neared, new clues about the merger between BP and the government explained a lot to those watching closely—especially those in the industry who questioned the tactics being used. Chu's science team—Tom Hunter, director of Sandia Labs, George Cooper, a retired professor from Berkeley, Richard Lawrence Garwin, a physicist, and Alexander Slocum, a mechanical engineering professor from MIT—is made up of researchers and teachers, primarily in nuclear physics or mechanical engineering. None of the team has a direct oil industry background. The Department of Energy Deepwater Response website that names this team says it has over 200 other advisers but fails to list anyone who has actually done any work in oil and gas.[45] Hopefully some have, but it's clear by some of the decisions that were made that if there are industry folks on the team, they don't have much influence.

Deciding to try to shut in the well with the capping stack was one of those decisions that an industry person would have strongly questioned. To do this is like flying an airplane into a storm without an instrument pilot's license. You might get away with it, but there's not much in the way of alternatives if something goes wrong—and dire consequences if it does. In mid-July, after shutting in the capping stack to stop the oil flow, some, mostly industry outsiders, have said, "You can't argue with success," and that shutting in the well worked. Maybe, but I question if it was good risk management. Most experienced people would say no, it was a risky procedure. Were they successful shutting in the well? Yes. Did they exercise poor risk management? Absolutely. Only after the capping stack was set did we find out that it was inadequately designed and only able to control the oil flow as an all-or-nothing proposition: Both BP and Admiral Allen told us that the stack must either shut in the well completely or flow all the oil into the Gulf after it was set.[46] So this stack, designed and engineered for two months before installation, was not designed sufficiently to contain only part of the flow and produce the rest to whatever capacity was available at the surface, a design defect that never made much sense.

Delaying the relief well, for weeks and weeks, especially with Boots & Coots's John Wright, the best relief well driller on the planet, in charge, was unbelievable. Had Wright been left alone, the well could have conceivably been killed in July. And safely. With the revelation of the US government's active management of this crisis, much of the confusion about decision making was cleared up. All the course changing, halting press briefings, and opaque, infrequent, and lukewarm "technical briefings" were now explained. BP was happy to appear to be taking the backseat to the government team, even though the government was dependent on information that came directly from BP. The reasons were simple: BP was off the hook from the moment the government took over its public presence, especially since its viability in the United States going forward was assured, and BP management would be more than happy to point the finger at the government if the whole thing blew up.

# EIGHT

# How Not to Save an Ecosystem: Dispersants, Skimming, and Booming

The most unsettling realization for people watching this tragedy unfold was BP's complete inability to control the massive and growing spill. When the Coast Guard announced a few days after the blowout that indeed the well was flowing and then gradually increased the estimates of flow rate, there was a panic among first responders, state and local authorities, and clearly the federal government. The problem, though, was that there were virtually no assets that could be brought to bear effectively. In fact, almost no work had been done on spill cleanup and dispersant technology since the last major disaster—when the oil tanker *Exxon Valdez* grounded on a reef and unleashed its cargo into Alaskan waters in 1989. It was every man and woman for themselves.

In the two decades since then, only a few environmental groups called for more research and development of cleanup technology. But interest waned in ensuing years, and dollars for research gradually diminished. Complacency took over. The oil industry appeared to have done their bit by forming the Marine Spill Response Corporation (MSRC), meeting their official obligation for contingency planning, and everything went back to normal. Normal, that is, until the next crisis. The Macondo blowout was that crisis, but by then it was way too late. The techniques of the 1980s wouldn't work in 2010 (of course they didn't work very well back then, either).

I fear that in coming years we are going to become painfully aware of the consequences of the methods used to deal with the oil that roared out of the Macondo well for 87 days in 2010. The response was an odd combination of obsolete 40-year-old techniques and untried, untested, unproven, and likely unwise application of massive quantities of toxic oil dispersants, not only on the surface of the Gulf but also subsea at 5,000 feet of water depth, an application never tried or even contemplated. We watched our televisions for weeks as dispersant was applied 24 hours a day, seven days a week, directly into the flow of oil coming from the severely damaged BOP. We watched as the Coast Guard and local authorities tried to get a handle on dispersant use and as BP ignored calls to stop or at least restrict its use, continuing to spray thousands of gallons a day from airplanes and boats.

In the early days of the crisis, BP applied dispersant from the air on the miles-long ribbons of reddish brown oil coming to surface and spreading over the Gulf, spraying almost indiscriminately before the EPA ordered a halt, of sorts, on May 10. In a letter to BP, it gave the company 24 hours to come up with a safer dispersant than the two they were using—Corexit 9500 and Corexit 9527.[1] In true form, BP essentially ignored the order, finally responding that the two Corexits were the only dispersants available in sufficient quantities to handle a spill of this magnitude. Environmentalists were in uproar, as were the Gulf states' environmental agencies, but BP persisted.[2] The EPA relented, allowing dispersants to be used in "rare cases."[3] But the Coast Guard ended up granting every exception request made by BP to apply dispersants, 74 times in 54 days.[4]

At the same time, BP, the MSRC, and jurisdictions all along the Gulf Coast were scrambling to get oil booms in place in the water to protect the sensitive shoreline. Booms are old technology, don't work anywhere but the calmest of water, and must be laid properly to divert any oil coming ashore to catch areas where it can be picked up. The booming that was done along the Gulf Coast, mostly by BP contractors, was laid improperly and really had little hope of holding back the oil from the shorelines, especially the marshes. The poor placement, along with inadequate manning, caused a good part of the boom that was exposed to wind and wave action to simply pull out of its anchors and wash uselessly ashore by the mile. As the magnitude of the crisis became clear, it was obvious that shoreline booming was being done more for the television cameras than to actually do any good. Where it really counted, it didn't work very well or, in many cases, at all.

A massive effort was begun to skim the oil off the surface, an old technique that works for light oils on flat water. Emulsified oil in open ocean, not so much. Even the government—which in August 2010 issued the hastily released, yet puzzling, Oil Budget report that supposedly analyzed where all the oil went—could document only that 10 percent of the estimated 4.1 million barrels spilled into the Gulf was either skimmed or burned (after skimming). That's pretty poor performance, especially after undertaking a massive effort of MSRC skimming ships and private boats, including an estimated 3,000 fishing boats hired through BP's Vessels of Opportunity program that paid local fishermen to skim oil and set boom.[5]

One lesson we should have learned in this tragedy is how poorly we are currently equipped to deal with a blowout of a deepwater well where the rig is lost and the wellhead is damaged. Let's look at these three inadequate efforts: dispersants, booming, and skimming.

## Flying Blind: The Chemical Carpet-Bombing Experiment

Early on in this disaster, BP made the decision, with the consent of the US government, to sacrifice the deepwater column in the ocean to save the shoreline and wetlands. They won't say it that way, of course, but that is exactly what they did. On April 29, BP began applying dispersants, first Corexit 9527 and later, after they ran out, Corexit 9500, directly into the flow of oil at the wellhead, a technique for application of dispersants that has never been tested or tried. About 1.5 million gallons were used, almost 800,000 at the wellhead, which again had never been done before.

By doing so, they broke up the oil into tiny droplets 5,000 feet below the surface of the Gulf.[6]

There are many unknowns and much disagreement in the scientific community about the results of that action. Everyone agrees that much of those dispersed droplets never rose to the surface but stayed subsurface. Much of the disagreement surrounds the "what happened next" scenario. Since it's pitch black at those depths, scientists considered whether the oil would biodegrade without sunlight. If so, would the process be slower? How long would it take? Is the concentration toxic?

If you listened to the NOAA and EPA scientists, you learned, initially, that much of the oil would evaporate quickly, and that fish metabolize whatever oil they ingest rapidly.[7] So seafood in the Gulf was suddenly perfectly safe. Independent scientists said, *Not so fast, it's not that easy.* The echoes of the EPA's happy talk after 9/11 sounded loudly.[8] Recall that on September 18, 2001, with the wreckage of the World Trade Center still burning and asbestos dust billowing into the air, Christie Whitman, then director of the EPA, declared the air and drinking water at ground zero to be safe. In her statement that day, Whitman said, "The public in these areas is not being exposed to excessive levels of asbestos or other harmful substances."[9] With hundreds of Ground Zero workers now suffering chronic lung disease and nerve damage, and some already dead, we now know how false and irresponsible that declaration was. It seems our government, under a different president, is doing it to us again.

Made by chemical company Nalco, Corexit 9500 and 9527 are widely used dispersants that are on the EPA approved list, although they have been banned for use in the UK since 1998 due to their toxic effects on certain shorelines.[10] We are familiar with the Corexit dispersants, since an earlier version, Corexit 9580, was used in Prince William Sound during the *Exxon Valdez* spill. Many scientists believe that its use there caused much of the long-term damage to the area's ecosystems in the years following the spill. These dispersants, though their exact composition is a trade secret, are made up of a combination of chemicals that essentially cling to drops of oil, breaking them into tiny droplets. We do now know that 9527 contains butoxyethanol, which is known to cause internal bleeding, and that 9500 actually contains petroleum distillates, which are solvents that break up oil. Essentially, when they used 9500, they put petroleum onto petroleum to break it up. But that's not the really bad news. Though the EPA disagrees, most of the environmental science community will assert that dispersant, mixed into oil, is more toxic than either substance by itself, hence the grave concern about the effects on sea life in the Gulf of Mexico.

Marine toxicologist Susan Shaw, director of the Marine Environmental Research Institute, along with other concerned scientists, began sounding the alarm about Corexit soon after BP began indiscriminate use. They began calling for BP to halt its heavy use of the product, especially after the company refused the EPA's May order and the agency capitulated. After the EPA issued a dispersant toxicity report in July 2010, Shaw issued a paper in response the next month, pointing out the weaknesses in the government report that helped lead the public to misinterpret the toxic effects of this dispersant. Shaw criticized the EPA's testing methodologies, asserting that they tested only the less toxic of the two dispersants, Corexit 9500, while ignoring the more toxic and bioaccumulative Corexit 9527. She also determined that the EPA's short-term toxicological tests were designed for coastal areas, not the deep sea, and that it ignored long-term effects on the ecosystem itself. That EPA report declared 9500 to be "moderately toxic" but also drew the conclusion that the oil–dispersant mix was not more toxic than oil alone. Shaw disagreed, as have other scientists, noting that this conclusion is simply not true, and that, in fact, these dispersants applied on the ocean floor could cause catastrophic damage to the marine ecosystems.[11-13]

Corexit was never intended to be used subsurface, much less at great depths. It was designed for surface spills, to disperse what oil had been missed by skimmers or oil booms. The application at the seafloor had never been tested, and its toxic effects on the water column are unknown today. The massive application of these dispersants was described by some as a giant experiment, "uncontrolled." Likewise, Dr. Ron Kendall, a wildlife toxicologist from Texas Tech who made multiple trips to the Gulf during this time, said of the use of these dispersants, "It's unbelievable. It's still unfolding. This is a catastrophe of enormous proportions. To me, this is the biggest environmental toxicology experiment we've ever conducted." [14]

## The Mystery of Deep-Sea Oil Plumes

While toxicity of dispersant use in the deepwater column had not been studied, indeed, the effect of a deepwater blowout had been. In 2000, a consortium of 23 companies, plus the Minerals Management Service, conducted a study called Deep Spill that studied the effects of oil and gas released in deep water, simulating a blowout. What they found was that, indeed, most of the oil stayed at a "neutral buoyancy point" in the water column determined by hydrostatic pressure, currents, and temperature. The study determined that gas separated from oil not far above the wellhead, and the oil formed clouds of droplets that hung in plumes below the surface. Only an average of about 8 percent of the oil made it to the surface, and the slicks were much thinner than from surface spills.[15]

These results were duplicated through simulation by researchers at the University of North Carolina in May 2010, finding that the vast majority of oil stayed below the surface in zones of colder temperatures and increased water density.[16] These results helped confirm the discovery of deep underwater oil plumes by the research ship *WeatherBird II* of the University of South Florida and NOAA's research ship the *Jefferson*. One plume was found to be 6 miles wide and 22 miles long, drifting at a depth of 3,300 feet. Testing of samples from the plume confirmed that the oil was definitely from the Macondo well.[17] Two days prior to the discovery of plumes, Tony Hayward had gone on record denying the presence of plumes, saying, "The oil is on the surface. There aren't any plumes."[18] That's pretty remarkable since BP was one of the participants in the Deep Spill study that predicted exactly what the research vessels were discovering. I guess nobody told Tony.

There are two primary issues with oil and dispersed oil in the water column. The first is how it immediately affects fish, shrimp, crab, and other sea life that is caught for consumption. As is understandable, there are varying opinions on the safety of seafood now being harvested. Oddly, it seems the usual roles here have been reversed in this case, with NOAA and EPA saying that the seafood is safe, and the *fishermen* not so sure. In August, as the government started reopening fishing areas, including the reopening of shrimping season, many fishermen voiced their concerns about the safety of the seafood, and some even refused to fish. In early August, a group of fishermen from all over the Gulf Coast met in Biloxi, Mississippi, to alert the public to their fears. Many reported strange behavior in the sea life, seeing bottom dwellers in shallow water. One Mississippi fisherman, Danny Ross, was quoted as saying, "Sea creatures that are normally bottom species are on top of the water due to the water column is so full of dispersants. The oxygen level is so low. A Horseshoe Crab, it's on top of the water trying to swim. We've never seen that in our life. We might not be biologists, but we know our waters. This is not normal."[19] Many of their fears about oxygen levels in the water were confirmed when a huge fish kill was discovered in late August off St. Bernard Parish, Louisiana. It was estimated by parish president Craig Taffaro that 5,000 to 15,000 fish were killed of all species: bottom dwellers, finfish, and crabs.[20]

The second issue is how these toxins affect sea life reproduction, internal organs, and food sources long-term—also a subject of debate. Susan Shaw, very concerned about long-term effects to the food chain, bases that concern on observations of the long-term effects of the *Exxon Valdez* spill and the use of Corexit in

Prince William Sound. She worries about a phenomenon called trophic cascade—a complicated collapse of an ecosystem starting at the bottom of the food chain. After the *Valdez* spill, says Shaw, the cascade was actually started with the die-off of algae and barnacles that held the substrata and the bottom of the sound in place. They were replaced with invasive species that were then ripped out with tide and wave action, making it less stable. With the disappearance of algae, the preferred food source, organisms at the bottom of the food chain, such as snails, were wiped out. Mussels that fed on those were wiped out, and the process moved up the food chain, eventually wiping out the entire population of sea ducks, estimated at 200,000 at the time of the spill. They have never returned. Other species, including herring, also never rebounded. Shaw is concerned with a similar collapse in the Gulf from the toxic effects of the oil in the deepwater column. It will take years to know the extent of the damage, since we still don't know this in Alaska after the *Valdez* spill.[21, 22]

The now famous Oil Budget, issued by several government agencies led by NOAA, attempted to account for the oil from the Macondo well. Of the estimated 4.9 million barrels of oil (the report ignored the natural gas), 800,000 barrels were said to be captured by containment to ships on the surface, never reaching the water; all but 1.3 million barrels were collected, skimmed, burned, "dissolved," or dispersed.[23] When the report was issued, only shortly after the oil stopped flowing, administration officials rushed to the microphones proclaiming the disaster over. Well, not in so many words, but they seemed happy nonetheless if the press got that impression. The government began opening fishing areas, and Jane Lubchenco, director of NOAA, who happens to be a respected marine biologist (at least up until this point), surprisingly said that seafood coming from the Gulf was perfectly safe. The problem was, nobody believed her or the rest of the government statements.

In a jovial White House briefing on August 4, Lubchenco, White House energy and climate adviser Carol Browner, Admiral Allen, and Press Secretary Robert Gibbs gleefully summarized the results of the Oil Budget to a confused press corps, complete with Gibbs adding misplaced humor by channeling Vanna White during the presentation.[24] The next day, the White House got exactly what it wanted when the *New York Times* printed the story with the headline, "U.S. Finds Most Oil from Spill Poses Little Additional Risk"[25]—which, of course, wasn't what the report actually said, but the Obama administration's political strategists high-fived that they had gotten the media to close the books on the ongoing disaster and move on to the next celebrity who was going to jail and the story of the mosque that wasn't really a mosque being planned on the sacred ground of Ground Zero that was actually an old Burlington Coat Factory almost three blocks away. The distraction and deception were complete, and the story fell off the front page and the television screens.

Soon after the Oil Budget report was issued, many in the scientific community took issue with it, claiming that the report had not, in fact, been peer-reviewed, as Carol Browner had claimed at the White House briefing. A University of Georgia report in August disputed the Oil Budget, saying that the media reported the data wrongly, assuming that "dispersed" meant "gone," when in fact it was not gone at all. The report concluded that 79 percent of the oil was still in the Gulf, much of it below the surface, and degrading at a much slower rate than claimed by the government.[26] Lubchenco further confused matters throughout this period, quibbling with the math in the Georgia report, but also acknowledging that "dispersed" actually doesn't mean "gone." At the same time, though, she reiterated that the government had proclaimed Gulf seafood safe to eat.[27]

At this writing, neither NOAA nor any other federal agency has provided the data or calculations that supported their hasty contentions, causing much concern in the scientific community. During a mid-August 2010 conference call with unhappy congressional staff members who were demanding the data in the report, NOAA said that it would be two months before they would make it available, since it was being "peer reviewed."[28, 29] The administration actually had successfully manipulated the media with happy talk that the oil spill risk had passed and were now stonewalling Congress to keep the actual data out of the public's hands until after the coming election.

At this writing, research continues. In mid-August 2010, the University of South Florida issued a preliminary report disputing the findings of the Oil Budget, claiming that researchers had actually found oil on the seafloor around and in DeSoto Canyon off the coast of Florida's panhandle and east of the Macondo well site. The canyon provides nutrient-rich water to important spawning grounds for commercial fish in the Gulf. Researchers found negative effects of oil and dispersants on microorganisms and plants in the area and noted toxicity in 39 percent of their samples.[30] Another report, from the Woods Hole Oceanographic Institution, noted that although the plumes they discovered were smaller than other researchers had reported, the biodegradation was much slower than that being reported by the government. As reported by the *New York Times*, Dr. Richard Camilli, the lead scientist in the study, said, "[Assuming that the physics they observed in June remained similar] . . . it's going to persist for quite a while before it finally dissipates or dilutes away."[31]

On August 24, 2010, Steve Chu's old employer, Lawrence Berkeley National Laboratory in Berkeley, California, issued a controversial report that made the surprising claim that a heretofore unknown and "new" microbe has been discovered in the deepwater of the Gulf that rapidly biodegraded subsurface oil plumes (you know, the ones that Tony Hayward claimed didn't exist). According to the study, this new microbe resides in very deep, cold water, eats oil, and doesn't deplete oxygen levels. Terry Hazen, an environmental ecologist who led the study, said in a statement, "Our findings show that the influx of oil profoundly altered the microbial community by significantly stimulating deep-sea psychrophilic (cold temperature) gamma-proteobacteria that are closely related to known petroleum-degrading microbes.

This enrichment of psychrophilic petroleum degraders with their rapid oil biodegradation rates appears to be one of the major mechanisms behind the rapid decline of the deepwater dispersed oil plume that has been observed." In other words, after pumping millions of barrels of oil and over a million gallons of dispersants into the deepwater column, a new microbe was created that rapidly ate all the oil. Some would also dispute the observations of a "rapid decline" in the plumes, as the studies from the universities of South Florida and Georgia pointed out.[32]

Hazen, who is well regarded in microbial research, has spoken out publicly against the use of dispersants, especially in the deepwater column. He has said that natural microbial processes would degrade the oil sufficiently without the use of these dispersants and that they should be used only when the oil is approaching sensitive areas.[33] This new study will certainly cause reevaluation of the use of dispersants and assumptions about the degradation of oil far below the surface; however, the validity of these findings will also be called into question, requiring considerable outside peer review for a couple of reasons. First, Berkeley National Lab is part of the Department of Energy now under Chu, and where Chu was previously director. Second, this study was supported by the Energy Biosciences Institute, a partnership among the DOE, the University of California, and the University of Illinois, funded with a $500 million grant from . . . BP.[34]

## Implications for Human Health

Besides the devastating effect of the oil and dispersants on the ecosystem and its inhabitants, one of the unintended consequences of this massive use of dispersants was the effect on the human beings either applying them or working in them. Early on in the response, before BP began supplying protective clothing, many cleanup workers reported ill effects such as nausea, vomiting, bad headaches, burning eyes, persistent coughs, sore throats, stuffy sinuses, dizziness, and skin rashes.[35] Marine toxicologist and author Riki Ott went to South Louisiana during the height of the spill, extensively covering the health crisis that was building even as the oil and dispersants poured into the Gulf. Having experienced the *Exxon Valdez* spill and cleanup firsthand, along with the community's attempt to recover from it, Ott has monitored the long-lasting effects of the *Valdez* spill for 20 years, writing two books, *Sound Truths and Corporate Myths* and *Not One Drop*. She has followed the progress of the *Valdez* cleanup workers and the chronic, life-threatening symptoms they continue to exhibit. In South Louisiana, workers are suffering many of the same symptoms that became chronic in Alaskan workers. According to an EPA air quality study she quoted in May, some toxic compounds in the air were at levels 100 times greater than that which causes health problems.[36] BP dismissed the reported health effects, claiming that respirators and other protective equipment weren't needed.[37]

Clearly the jury is still out about the long-term effects of this massive experiment on one of our most important ecosystems. In the meantime, confidence in our government agencies has been undermined to the point of hostility between those who make their living in the Gulf and those who govern them.

## Booming and Skimming

There's actually not a lot to say about booming and skimming, except that neither works well unless conditions are perfect for it. If the use of dispersants is the weird science experiment part of oil spill remediation, skimming and booming is the Neanderthal— the technology and techniques of both haven't advanced much in the last 40 years, and neither technique is really appropriate for deepwater blowouts, as we witnessed over that last four months. The main problem with both techniques is that for either to work, you must have virtually flat water. Wave action is extremely detrimental, because oil and water mix into an emulsion and/or the oil stays below the surface. Neither works in surf, nor when the wind is blowing, nor when seas are much more than a light chop. The reason for booming and skimming is mainly for the television cameras to have something to record, and, in this particular case, to give BP a reason to keep out-of-work fishermen employed doing something while the oil fouled their fishing grounds.

Proper oil booming is not as easy as it looks. Boom is usually an inflatable tube that floats in the water with a skirt that hangs below it, and the idea is that it collects oil on the surface so it can be skimmed or burned. The boom is laid in sections, with anchors on each end to hold it in place. Proper booming, according to my booming expert friend, involves actually channeling the oil to collection points so it can be gathered. It's not intended to be laid mile upon mile in a straight line like a wall, which won't work, especially with any kind of weather. When boom is laid properly to protect a shoreline, laid at an angle to wind and tide, it will actually funnel oil to a point where it can be easily collected. When it's laid improperly, as we have seen in the hours of aerial footage on television, the oil washes back and forth across the boom, then eventually over it or under it as waves hit and tides change. The absorbent boom that is usually laid behind the inflatable boom will catch some of the oil but will quickly become saturated and will allow oil onto the shore or marsh grass with the change in tides. In short, shoreline booming is simply an ineffective technique to catch oil from massive spills.

Early in the response, when the Coast Guard, BP, and local authorities were grappling with the massive spill, many Louisiana parishes took matters into their own hands, with the blessing of Bobby Jindal, Louisiana's governor. Against the orders of the federal government, and the advice of marine scientists, Gulf Coast parishes began their own booming program not with inflatable booms, but by dredging the bottom of wetlands and building huge sand berms that would supposedly hold off incoming oil.[38] The area they were dredging was habitat for sea turtles, and a number were killed during the process. The berms, after being built, blocked access to the beach for these sea turtles and other shoreline sea life. The federal government halted the berm construction on June 23, igniting a furor from the parish presidents and the governor.[39] The feds relented soon after and let the construction continue, all funded by BP. In the ensuing months, many of the berms have been washed away by wave action and their effectiveness questioned by scientists. Some scientists are concerned that the berms were not very effective against the oil, were dangerous to sea life, and in fact may have sped up coastal erosion by loosening the sediments in front of them.[40]

There have been many attempts to improve traditional skimming, but the techniques are very limited. In open water, efficient mechanical skimming is almost impossible and collects only part of the oil that is on the surface. A common technique is to vacuum small amounts of oil off the surface with hoses after floating boom contains it, but results are usually incomplete. Another technique is to gather oil with floating boom between boats, collecting it at a central barge or boat. Those boats can use what are called weir skimmers, which basically flow the oil–water mixture right at the surface of the water into a containment vessel in an attempt to take just the oil. They also use absorbent materials to gather the oil, and sometimes belt skimmers that pull the oil from the surface.

On flat water, drum skimmers are used, which rotate right at the surface. Oil adheres to the drum, which is then wiped with a rubber blade running along the surface, where upon the oil runs into containment. The most effective shallow-water collection is with absorbent materials, applied by hand. Disposal of the contaminated materials then becomes the huge challenge.

It's interesting that the only dollars spent on skimming development have been by private individuals and companies. One such technique that was encouraged by some television pundits, but ended up a total failure, was the A Whale, a converted supertanker owned by a Taiwanese shipping company. The ship had huge slots in its sides to act as weir skimmers designed to be large enough to collect huge volumes of oil–water mixture in open ocean. After a week of sea trials, the Coast Guard rejected the ship because it was so large that it could not maneuver well enough to collect the ribbons of oil and could not collect any of the oil just below the surface.[41] The actor Kevin Costner also made a serious effort to develop skimming technology, inspired by the *Valdez* disaster. He formed his own company, Ocean Therapy Solutions, and invested $20 million of his own money to develop a centrifugal separating system that pulls oil–water mixture into a vessel, separating the oil and returning the clean water to the ocean. After a demonstration, BP purchased 32 of Costner's units. The questions, however, regarding all of these surface techniques are how much oil is actually removed, and whether they're able to remove large quantities. The vast majority of skimmers are small scale and effective only on flat water.[42]

The bottom line is that when it comes to removing oil from water, nothing currently available works very well, and millions of research dollars need to be spent to prevent another catastrophe like this one.

The cumulative health effects of the oil are bad enough; add dispersants, and you have not only an ecological catastrophe, but also a human health tragedy. The oil and gas industry, for decades, living in its own world, had convinced itself that there was no need to invest in better cleanup technologies since there was never going to be a spill that required a cleanup. That myth, so publicly destroyed, has exposed the critical need to develop containment and cleanup technology if we are to drill the deep-water safely.

# NINE

# Top Cap, Top Hat, Top Kill, Capping Stack: Making It Up as We Go Along

Of the many failures that occurred before and after the blowout of BP's Macondo well, one was glaringly apparent in the very first hours. BP, Transocean, the oil and gas industry, the Coast Guard, and the US government were all woefully unequipped to deal with not only the blowout and burning rig, but also the environmental catastrophe that followed. The lack of preparation was as eye opening as it was shocking. After the searing images of the rig sinking, many took some small comfort, at least, from hearing that the well had somehow bridged over and was not flowing. No one in the industry believed that, but that was the initial story from the Coast Guard, told by Rear Admiral Mary Landry, with BP standing mute, as became a habit over the next three months, letting the Coast Guard do the dirty work. A few days later, the nightmare news broke that the well was actually flowing in 5,000 feet of water. Estimates worsened every day as the catastrophe unfolded.

As we know, the Coast Guard first estimated the well was flowing 1,000 barrels per day. In about a week, they raised it to 5,000 barrels per day, even as scientists were estimating much higher rates. But the government stubbornly clung to the 5,000-barrel-per-day estimate for five weeks. By mid-May, scientists estimated the flow at 25,000 to the 70,000 barrels per day estimated by Wereley—who used technique called particle image velocimetry, analyzing videos of the flowing well.[1] On May 20, BP reluctantly made the video feed of the blown-out well public. That feed rapidly became the obsession of millions watching in fascination around the clock; one group of users at Daily Kos organized a vigil of sorts, calling itself Gulf Watchers, with at least one person reporting goings-on with the various ROVs and the ships on site during the crisis. I found their observations not only very helpful, but comforting, in their own way, during some of my own late-night vigils watching the oil roar into the Gulf and the continuous efforts to contain it.

As people watched the video feeds and satellite images of the spreading oil slick on the Gulf of Mexico's surface, estimates of flow began to climb. No one in the industry ever believed the flow was less than 20,000 barrels per day. While BP remained mum, the Unified Command under Admiral Thad Allen, the national incident commander, used the 5,000-barrel-a-day figure until the Flow Rate Technical Group, made up of five teams of scientists including Secretary of Energy Steve Chu and represented by USGS Director Marcia McNutt, announced on June 10 that the flow was indeed not 5,000 barrels per day, but as much as 40,000. McNutt insisted that this was an early estimate, and the work wasn't complete.[2] It didn't matter, though; the news media went into overdrive, and the news of the new rate spread like, well, an oil slick. Concern on the Gulf Coast turned into panic as word got out. Suddenly, it sank into our collective consciousness that this was easily the largest environmental catastrophe in the history of the United States, dwarfing the *Exxon Valdez*.

From the beginning, we watched BP struggle with this monster as the rest of the industry stood ready to help, but also not knowing what to do. Shell sent boats and offered its offices in Robert, Louisiana, as a headquarters for the newly established Unified Command, made up of a whole myriad of federal agencies, including most of the alphabet agencies like the Department of Homeland Security, Department of Energy, Department of the Interior, MMS (now BOEMRE), EPA, Department of Defense, FDA, USDA, NOAA, USGS, and, of course, BP and Transocean. With Admiral Allen in command, daily press conferences were held there as everyone tried to get organized to form a response.

As the weeks dragged on, each time BP tried something to stop, or even slow, the gushing well, someone from the government or the company would say, "This has never been done before," or, "We've never tried this before," and so on. It was clear . . . no one had ever even anticipated a deepwater well blowing out where the BOP and all other safety systems failed. And no one had the slightest clue what to do next.

Hundreds of billions of dollars have been spent over the last forty years by the oil and gas industry to develop very sophisticated exploration and drilling technologies. So, why has little been spent on the design of cleanup or containment methods for a blown-out offshore well? And why has zero been invested in deepwater containment and recovery? Because the industry actually believed that whiz-bang technology (more of that PFM) made a blowout and spill "impossible." If you didn't believe it, all you had to do was ask a deepwater operator, or read their offshore development plans filed with the MMS.

In the days after the 1989 *Exxon Valdez* spill, there was a public outcry for stronger regulation and penalties against companies that spill oil. Among the provisions in the Oil Pollution Act of 1990, Congress's response to the *Valdez*, was one requiring offshore operators to provide their own spill response plans and resources. The industry pooled resources then and started the nonprofit Marine Preservation Association (MPA), which in turn funded the Marine Spill Response Corporation, intended to respond to offshore oil spills on behalf of its members. The convoluted structure, of course, was designed by oil industry lawyers to shield companies who used the MSRC from liability for its skimming operations. Currently costing the industry about $80 million per year, it was intended to lead the charge in responding to this spill. The problem, though, was that the MSRC was never equipped for or dedicated to responding to deepwater spills. Even Steve Benz, MSRC's CEO, admitted their inability to respond to the spill, saying to the *Washington Post*, "Should the industry's capacity have been greater than it is? That's a fair question."[3] To add insult to injury, a for-profit competitor came on the scene in 1992, the National Response Corporation (NRC), and siphoned off MSRC participants by offering the service at a lower cost. The result? MSRC lowered its cost by cutting research in improving oil cleanup technology. The "free market" reigns again, and research suffers, because there's just no profit in research for cleaning up a deepwater oil spill that's not going to happen anyway, or so everyone thought. So the lame response of 2010 was actually predestined in the early 1990s by the crippling of the very organizations that were established to stop catastrophic spills.

## The Cost of Complacency

The complacency and overconfidence in the industry that "this just couldn't happen" really played out on the *Horizon* itself. Not only were several of Transocean's alarm and shutdown devices either inhibited or turned off altogether, but the very design of the rig proved part of its undoing. The majority of the damage from the rig floor blast went through the crew quarters and galley, where people are most vulnerable. The explosions in the engine rooms damaged and cut off access to the aft lifeboat station and started the fire that likely caused the blowing well to light off. With the general alarms inhibited, the crew got no warning before the blasts. Many witnesses have stated that they heard no alarms even as they smelled gas, and that only some heard announcements on the PA system. The magnitude of the blast and the structural damage rendered the muster and firefighting plans virtually useless. They couldn't fight the fire anyway, because the firefighting system was tied to the deck engine generators that were inoperable. The standby generator also wouldn't start. The rig was dying a quick death. In the panic during the moments after the blast, it was literally every man and woman for themselves. It was pure luck of timing, circumstance, weather, and personal bravery on the part of several crewmembers that prevented this disaster from having many more casualties.

On an offshore drilling rig, at any one time, almost half of the crew is off duty and asleep or relaxing in crew quarters, especially when it is hours before they come on duty. Crew change generally happens at midnight and noon, with most crew working 12-hour tours (pronounced *towers*), or shifts, and the hand-off between crews starts about 30 minutes prior. On the night of the blast, much of the crew was apparently short-changing, meaning they were shifting earlier than normal. This usually occurs when they switch tours. In the offshore Gulf of Mexico, many of the crews work hitches of 21 days on and 21 days off. Some work 14 and 14. During their first week out, they will often, depending on their individual job, work the tour beginning at midnight. During the second week, they switch to days, and so on. On the day that they switch, they will sometimes short-change to make the transition easier. This system makes sure that everyone over time is treated equally, working both days and nights. On the night of the blowout, many of the crew were short-changing. So, at the time of the blast many of the crew who were off tour had not yet settled into their quarters for the night, especially since there had been a meeting with the BP and Transocean executives at 7 p.m. that evening. Had this been a normal crew change night, there could have been many more people in the crew quarters, increasing the chance for casualties from the blast that destroyed that part of the rig.

In the chaos after the blast, many of the crew were left on their own. Even the handheld radios stopped working. Survival instincts kicked in, and many jumped the 75 feet to the water rather than risk trying to get to a lifeboat station. At the forward lifeboat station, there was so much confusion and panic that muster lists were abandoned, as were some of the crew members, including the captain and those working on the bridge. Each lifeboat was capable of holding 75 people, so two lifeboats could carry the entire crew, but they didn't wait for everyone, so both were lowered with open seats as people inside panicked and demanded they depart. When the captain finally gave the abandon-ship order, it was too late for the bridge crew and one of the most severely injured, Wyman Wheeler, to get on a lifeboat. Luckily, the remaining survivors were able to get a life raft deployed and get Wheeler off with a few others in the boat while the rest jumped.

Had this blowout occurred in wintertime, or during a storm, many more crew would have died from exposure or just not been found. Had the *Damon Bankston* not been on station with a fast rescue craft and very able crew, again, the death toll would have been greater. By luck and the courage of several individuals that night, many lives were saved. But the blowout and ensuing chaos exposed terrible weaknesses in training and rig design that must be corrected.

With the entire drilling floor and most of the aft portion of the rig ablaze, it didn't take long for the 240-foot derrick to melt and collapse. The rig was surrounded with offshore vessels and fireboats pouring millions of gallons of seawater onto the rig, trying in vain to suppress the fire and cool the red-hot steel. The rig started listing badly and sank on April 22. The Coast Guard did not have a firefighting boat on scene and now admits to not having had a fire marshal coordinating the fireboats. There is now wide speculation that pouring the millions of gallons of seawater onto the rig actually caused it to sink, or at least contributed to it sinking. Had it not sunk so soon, the fire would have continued for some unknown amount of time but could have substantially reduced the amount of oil spilled in the early days of the blowout because it was all being burned. The Coast Guard is now investigating the firefighting that went on by the private boats and how it may have contributed to the sinking.

The rig settled to the bottom on its side 1,300 feet to the northwest of the wellhead. The 5,000-foot riser parted in several places as it was pulled under, torquing the BOP and wellhead with tremendous stress as it fell. What remained of the riser attached to the BOP was strewn across the ocean floor, with the buoyant sections eerily floating as high as 1,500 above the wreck. The *Deepwater Horizon*, along with the eleven missing workers, was dead. But the environmental catastrophe had just begun, and no one knew what to do next.

A deepwater blowout was not supposed to happen. If it did, the BOP was supposed to stop it. Every time. BP, and indeed the world, was soon to learn that this didn't happen, which would require a subsea engineering and operations effort never before witnessed.

## The ROVs

Remotely operated vehicles are the eyes and hands of operators in the deepwater—one of those marvels of technology that actually do work, and work well. Humans certainly can't survive at the depths where deepwater wellheads sit, so the ROVs do their work for them. They are surprisingly nimble, but much stronger than humans and able to manage hand tools, saws, and screwdrivers. Operating an ROV takes many hours of training and experience, and a pilot's skills are highly valued. It became evident in the early weeks of watching the streaming video that these skills were going to be put to the test in the coming weeks and months. And they were.

BP's first efforts to control the well were to shut in the blow-out preventer was by using ROV intervention, meaning that they could send commands by the undersea robots through a panel on the BOP stack itself . . . but nothing on the panel worked. So much for fail-safe technology. BP spent a day with an ROV trying to close a variable-bore ram on the BOP, only to find out afterward that the ram they were trying to close was the useless converted test ram. That was a mistake made thanks to the drawings supplied by Transocean—the drawings that didn't include modifications made to the BOP. Everybody was scrambling trying to figure out what was going on. They were eventually able to fire all six rams, but the damage to the internal components of the BOP, plus the presence of two pieces of drill pipe inside the stack, made sealing impossible. Still, no one could explain how the rams that had worked perfectly and held pressure just hours before the blowout completely failed when they were needed for real. The BOP wasn't closed, and would never close.[4]

On May 3, 2010, there was a brief glimmer of hope when Jeff Childs, a BP exec and deputy incident commander, announced at a briefing with Senator Richard Shelby that they had successfully closed the annular preventer on the top of the BOP stack, saying, "We've significantly cut the flow through the pipe." BP refuted Childs's claim a few hours later, saying, in a blunt statement, "BP would like to clarify that, contrary to some media reports, the actions it has taken to date on the blow out preventer have not resulted in any observed reduction in the rate of flow of oil from the MC252 well."[5] BP, like everyone else, was at a loss. The BOP, the last line of defense, was useless, sitting on top of a roaring well on the bottom of the ocean. BP had to do something, so they started making things up as we all went along for the ride.

## The Containment Dome

The next steps BP took never made much sense, and I believe they were more for show—to look like they were doing something while they were trying to come up with a real plan. The silliest contraption they built was a giant containment structure. It was a massive thing, over 40 feet tall, that was intended to be set over one of the three leaks in the wrecked riser to supposedly capture the oil and flow it out through a riser connected to the *Discoverer Enterprise,* an 835-foot drillship that had been moved into place.[6]

It took 15 hours to lower the structure over the leak, and the results were almost instantaneous: failure.[7] BP was forced to quickly halt its effort to capture the torrent of flowing oil due to hydrates that clogged the containment vessel. Hydrates are often formed when methane hits water under the right conditions, a common problem in offshore risers and pipelines. In this case, as oil and gas flowed from the blowout well and hit the cold seawater, water molecules essentially trapped methane molecules without chemically bonding with them. This trapping of methane formed hydrate crystals, which built up so quickly that the containment vessel, a large concrete-and-steel box attached to a riser that was designed to capture the majority of oil escaping from the uncontrolled well, was indeed clogged and actually began to float off the seafloor. BP threw in the towel, lifted the vessel off the leaking riser, and set it aside.[8]

## The Top Hat

At the time of the containment structure failure, BP announced that they were preparing to run a new, much smaller dome over the riser leak, calling it a "top hat," and hoped to have it on bottom at the end of the week. It was only 5 feet in diameter, much smaller than the first dome, designed with the idea of reducing the volume of seawater within the vessel, thus lowering the chances for the formation of hydrates. They planned to set it already connected to Transocean's *Discoverer Enterprise*, so that it could immediately begin to gather oil rather than sitting unconnected for a period of time with oil flowing into it.[9] Another key difference between this new device and the failed containment dome was that the top hat had a double riser running between it and the drillship that would be pumping methanol down the outside pipe and producing oil up the inner pipe. The methanol would be used to prevent hydrate and ice formation. This plan was not implemented. I believe that BP already knew the flow from the well was much greater than the top hat could ever conceivably handle.

## The Riser Insertion Tool

The next contraption that didn't work, at least in terms of capturing a good bit of the oil, was the riser insertion tool. It was an elaborate bit of technology that was essentially a straw stuck into the end of the wrecked riser to take oil to the surface. Again, methanol was injected into the stream to prevent hydrate formation, but nitrogen was also used to jet the oil to surface up the riser to the *Enterprise*. BP initially announced that they expected to collect up to 85 percent of the flow with the RIT, but in the end it captured only about 2,000 barrels per day at best. One day they were actually able to capture, for a short time, at a rate of 5,000 barrels per day, but they were unable to maintain it. After the partial success of the RIT, even BP began to reluctantly admit that the flow was greater than 5,000 barrels per day.[10]

## The Top Kill

After the RIT proved inadequate, BP announced the next idea to kill the well—and the first one to make any sense. The plan was called the top kill; it included a technique they called the junk shot. These are new terms to people outside the industry but familiar to those on the inside. BP believed that the partially closed rams in the BOP created a restriction, making a junk shot possible. They tested various combinations of "junk" made up of various sizes of rubber balls, rope, ground rubber, and other materials (even golf balls) to pump into the well through the kill and choke valves at the bottom of the stack. The hope was that these would hang up in the BOP long enough to stop the flow; if that worked, they would then pump kill mud from those same valves into the well to stop the flow. This method of top kill—called bullheading—requires a lot of pressure to get started, but once the hydrostatic pressure of the column overtakes the formation, it takes less to get it to bottom. The plan wasn't as crazy as everyone thought. In fact, pumping junk like this is an old oil field trick. Back in the day, one of the most reliable ways to find a drill pipe leak was to drop a bunch of chopped-up cotton rope into the pipe, then pump it down the well. When the pump pressure jumped, you stopped and pulled the drill pipe. When you found rope hanging on the outside of the pipe as you pulled it, you found your leak, having been plugged with the pieces of rope.

Even though BP announced their intent to try the top kill after the containment dome and RIT failed, it still took weeks of preparation before they finally started the procedure on the Thursday before Memorial Day. They set a manifold next to the wellhead two weeks in advance but then spent the better part of a week with ROVs actually having to cut off the old kill and choke lines from the BOP and installing connectors they could attach to the manifold. During these weeks of delays, I began to believe that they were intentionally going slow, hoping that the flow from the well was slowing. Typically, these big deepwater wells come on strongly, as much as 40,000 to 50,000 barrels per day, but decline very quickly as flow is carefully controlled to maintain productivity. Here, with uncontrolled flow up open casing, it's very likely that it did come on at those levels, or well above them, but declined precipitously as the days went on. We now know that indeed this was the case, but we will never know if BP was counting on that as preparations for the kill continued at what looked like a snail's pace. During this time, BP finally announced they could read the pressures inside the BOP, though they didn't disclose them for weeks after.

Finally, on May 26, BP started the top kill. Information from the company came painfully slowly as we watched thousands of barrels of mud come roaring out of the widening cracks in the bent-over riser at the top of the BOP on our televisions. Wednesday night passed; time dragged on until Thursday afternoon, when word got out that they had actually stopped the procedure for 16 hours without saying anything besides the "everything is going according to plan" mantra that had become so familiar. Millions who had been watching the video feeds for hours were outraged, including some politicians. Trying to do damage control, Doug Suttles appeared on CNN that evening, actually apologizing for BP not being forthcoming with information. However, the damage was done.[11] The trust gap, which had been widening due to BP's tight control of its messaging and almost continuous gaffes by Tony Hayward, still their CEO at the time, became insurmountable. From the outside, it looked like BP just didn't get it. From the inside, I think they were overwhelmed with the scale of the crisis; their lawyers were likely trying to shut down all communication to protect their clients' interests.

On Friday the 28th, Admiral Allen made a public statement that they had stopped the flow of oil with mud, fueling a premature media frenzy that the top kill had been successful.[12] He later backed off that assertion when BP said they would know that the top kill had worked only if the well stopped flowing. BP continued with its messaging of "We won't know for 24 to 48 hours" even though it had already been 48 hours. It became obvious on the morning of the 29th that the top kill had not worked. I heard from an industry source that the top kill had failed, but that they were going to keep pumping on it over the weekend rather than immediately announce it. BP officials even gave a tepid briefing that morning, staying very noncommittal about the success of the top kill. Just hours later, they abruptly announced that the top kill had failed.

That evening, Suttles said that they were disappointed the procedure hadn't worked, and the decision to end it had been made jointly between the government and BP. Over a month after the failed top kill, the *Wall Street Journal* published a story that the government had ordered the sudden halt to the procedure on Saturday due to a "malfunctioning disk" 1,000 feet below the seafloor.[13] There were three subsurface rupture disk devices installed in the 16-inch casing in the well at about 1,000 feet, 3,300 feet, and 4,500 feet below the mudline; they were designed to relieve pressure from the outside of that casing to prevent collapse in the event of a pressure buildup. It is possible that one of them failed or the entire tool failed, but the source was never disclosed; the story remained unconfirmed by other news agencies.

The top kill failed because they just couldn't build enough back pressure in the top of the BOP and bent-over riser to overcome this warhorse of a well. They pumped mud at up to 80 barrels a minute and slowed it, but they couldn't get enough mud in the well bore deep enough to build hydrostatic pressure. The "junk" they pumped in to bridge over the partially closed BOP and damaged riser didn't create enough restriction to help. It was time to go to the next step.

## The LMRP Cap

The next step to contain the flow was immediately announced by BP. They'd had so little confidence in the top kill procedure that they already had this and several other similar devices already on the seafloor ready to go when the top kill failed. The lower marine riser package cap was essentially a containment dome designed to fit over the riser, on top of the LMRP, which is the top component of the blowout preventer. Before they could land the cap on top of the LMRP, they had to cut the bent riser that was still attached. For weeks prior, I kept asking myself, *Why don't they just unbolt the riser flange and get the whole thing off?* Instead, they decided to cut off the riser with a sophisticated diamond wire saw that would mount directly to the wellhead and cut through the 21-inch pipe. The complication was, however, that there was still drill pipe inside the BOP and the bent-over riser that had been there when the blowout occurred. What we didn't realize, however, was that there were actually two pieces of drill pipe. That fact was discovered earlier by a science team, led by Steve Chu, that had used gamma ray imaging to inspect the BOP. When Chu told BP about the second pipe inside the BOP, they said it was impossible. Apparently it wasn't.[14]

After struggling for days to get the riser cut away from the BOP to take weight off it, they finally got it cut with a hydraulic shearing machine on the evening of June 1. I had been watching the live feeds all day, watching them struggle with the riser and the shear without much luck. That evening, by serendipitous timing, I was appearing live on MSNBC's *Countdown with Keith*

*Olbermann*, talking with him about the installation of the LMRP cap, when they finally cut into the riser. Not knowing what was going on, I finished my appearance and was getting ready to leave the studio when an engineer tracked me down and held out a phone. It was the producer, asking me if I had seen the feed in the last few minutes. I opened up my laptop and called up the feeds, and sure enough, they were cutting the riser. They immediately put me back in the chair in the studio, and I was able to describe to Keith live on the air what was going on.

The next step in the process was to cut away the rest of the riser from the very top of the BOP assembly. I didn't understand at the time why they just didn't just unbolt the flange right below it; in retrospect, they must have been concerned that the drill pipe on the inside might get in the way of bolting a new flange on top of the old. With two pieces of drill pipe on the inside, surely they would have known that it would have been almost impossible to cut because the loose pieces would jam the saw. The saw they used was essentially a very sophisticated band saw that clamps onto the riser connector flange and then slices the steel with a diamond-encrusted wire, leaving a precise, smooth cut. Sure enough, about two-thirds of the way through the riser the wire jammed, probably stuck in the pipe. To get it finally cut, they had to enlist the services of the hydraulic shear again, ripping the riser the rest of the way off. The rough cut precluded the use of the tighter-fitting LMRP cap, but BP had also designed a larger one that just set over the ragged top of the pipe. It was finally set on June 4, and BP then struggled with it for several days until they got it functioning. At its peak, the new cap, along with the converted choke and kill line, recovered about 25,000 barrels per day, less than half of the flow.[15]

## The Capping Stack

Amazingly, the capping stack was one of the first ideas proposed by BP's engineers to control the well and was floated as early as four days after the blowout. The stack was engineered, designed, and constructed out of virtually all standard Cameron subsea components, with the exception of the spool, or connection, between the new stack and the old BOP. The stack itself was ready and sitting on the dock in Venice, Louisiana, for some weeks prior to installation, and I never could figure out why BP seemed to be dragging their feet on it. This new capping stack was made up of a double ram cavity and a single cavity with both choke and kill lines installed with remote-controlled valves. On the bottom of the small stack was a Cameron HC latching connector that connected it to the transition spool below, and a latching collet connector looking up that allowed a riser from the surface to be attached. A variable choke attached on the choke valve would allow gradual closing or opening of the well.

The weakness in this solution, though, was not the new stack itself. It was the connection to the old blowout preventer stack. On top of the LMRP, which is the top component of the stack, are two connectors. One is the flex joint that allows the riser to move with the ocean currents or with the rig as it floats on the surface; the other is the riser connector that attaches the riser to the BOP stack itself. Being above the BOP, these components are not designed to withstand wellhead pressures that can go as high as 10,000 to 15,000 psi. Also, since the flex joint is designed to move, it is not a stable platform on which to set a new 150,000-pound capping stack. I was never sure why they couldn't remove the flex joint flange from the top of the LMRP, or why they didn't just remove the LMRP, which had a hydraulic latching connector similar to the one they were going to put on top. It's possible that there was some undisclosed damage in the connector; this was the same connector that should have released when Chris Pleasant, the subsea supervisor, hit the EDS button all those long weeks ago. Because the flex joint was so compromised, BP engineers had to actually straighten it with a hydraulic jack and then shore it up with wedges to keep it from bending when the stack was put in place. Even stabilized, it was not designed to hold the kind of pressures possible from shutting this well in, but that is exactly what they were about to do.

During the first week of July, BP kept talking about the construction of an elaborate set of four freestanding risers that would send up to 80,000 barrels of oil per day from the capping stack to containment ships. The risers were ingenious. Because they were installed below the surface of the water, they could remain in place even if a tropical storm required evacuation of the well site. Flexible hoses (which are easily installed) would connect the risers to the ships. On July 8, Admiral Allen sent a letter to Bob Dudley at BP demanding a timeline for getting "complete containment," including the capping stack and the surface vessels. In the letter, he also mentioned the possibility of "transitioning from full collection from the capping stack to shutting in the well using the capping stack." He ordered them to present a plan by the next day after a meeting with the government science team.[16] Dudley responded the next day with a detailed plan for installing the capping stack operations. In the plan, there was a "Close BOP & Monitor Pressure" line that indicated the test would be less than 24 hours. That was followed by "Implement Well Management Plan" that ran off the timeline.[17] Admiral Allen ordered BP to install the stack. The capping stack was set on Sunday, July 12.

## Well Integrity Test

After the capping stack was set, BP surprised everyone by announcing that now that they had a "capping stack" set, they were not going to actually hook up all the ships they had on station to collect the oil. Rather, they were going to run a well integrity test. My first reaction when I heard the announcement was, *What? Well integrity test?* I didn't recall even a single mention of a well integrity test up to that point. This surprised me, as I consider this test to be probably the most significant (and risky) operation BP had conducted since the failed top kill procedure during Memorial Day weekend (prior to the static kill in August, which was more risky). Digging back through the correspondence, I found this sentence in Admiral Allen's letter of July 8 to Bob Dudley: "BP has also presented general plans for implementing a capping stack procedure for complete collection of oil with the possibility of shutting in the well depending on results of tests to determine well integrity." [18] That's it, the only mention of well integrity. With the announcement, though, Admiral Allen signed off on a 6- to 48-hour well integrity test.

Even though the stack had been set for three days, they hadn't actually hooked up the two new valves to be ready to collect the oil and didn't intend to. Looking back at the timelines in Bob Dudley's July 9 letter, it didn't actually say they would. It said only "Implement Well Management Plan." [19] They had done it. BP had just pulled off the one thing they needed to do: shut in the well permanently before the flow was actually measured. And they did it with an order from the US government.

More alarmingly, they stopped drilling the relief well, which by now was only 4 feet away laterally from the blowout well. They had only about 30 more feet to drill before they got to casing point, the depth for the last string of pipe. *Thirty feet, and they stopped.* They sat circulating on bottom at 17,840 feet for days waiting on this test, wasting great weather to get the relief well finished—really the only solution to this blown-out well.[20]

On July 15, BP shut in the well with the capping stack and began the well integrity test. In his briefing prior to shutting in, Allen described the parameters of the test, announcing that they were going to take the stack, including the flex joint in between the stacks, to as high as 9,000 psi for up to 48 hours. He said that 8,000 to 9,000 psi would show strong integrity, 6,000 to 8,000 psi would be ambiguous, and below 6,000 psi would indicate a leak. When pressure was finally reported the next day, it was virtually level at 6,700, at the lower end of the reported ambiguity range, looking like there was leak-off from the wellbore down below. Since there was 1,200 feet of open hole from the bottom of the 9⅞-inch liner outside the 7-inch production casing to TD at about 18,300 feet, that made some sense. That's not to mention possible casing damage up the hole.[21]

Think of this well like a garden hose with a nozzle on the end. As long as the nozzle is open, the hose looks fine. As soon as you close the nozzle, the hose will leak through any pinholes or around the faucet as pressure builds inside. In his statement that day, Admiral Allen indicated they were probably going to go back to containment, which meant they'd be flowing the well to the various ships they had on station, but that never happened. Six hours became 12, then 24, then 48; for the next several days, Allen kept giving BP authority to keep the well shut in. Pressure gradually built, but almost imperceptibly. Speculation started that pressures were low due to depletion from all the production. That was likely, especially since the flow rate estimate was now up to 60,000 barrels per day. As pressure gradually built, BP and Allen declared they had well integrity. Maybe, but their main goal was to keep that well shut in. Oh, and the flex joint? It started leaking on the first day.

On July 17, BP started waffling about their intentions. Obviously, relations were already tense between the admiral and the BP guys. Allen ordered BP to open the well to containment at the end of the current period. On the 17th and 18th, Suttles and BP Spokesperson Kent Wells both said that in order to flow the well back to the surface, oil would flow into the Gulf for at least three days before they could contain it. They started the new mantra, "Nobody wants to see any more oil going into the Gulf."[22, 23] Of course, nobody asked why it would have to flow into the Gulf since they had containment lines already attached to the blowout preventer and to the surface. It was a showdown. BP was going to force Admiral Allen to order them to reopen the well. He blinked, and the well stayed shut in on continuous "well integrity testing."

## Static Kill

BP rolled out its next surprise on July 19. During Wells's afternoon briefing, he suddenly announced they were studying a new procedure called the "static kill."[24] The static kill is essentially a top kill, this time with a closed wellhead to provide back pressure. The procedure, bullheading, is to pump mud into a well from the top rather than circulating it in through a pipe run inside the well. This can work in a static well if you can get mud to start into the well and don't exceed the working pressure of the wellhead. As more mud is pumped in, hydrostatic pressure from the heavier mud will build, pushing the oil and gas back down the well and into the formation. It differs from the top kill that BP tried in that the well is shut in (thus, static), and they now have a closed system rather than the leaky riser they dealt with before. The problems with this procedure are the integrity of the wellbore—here, one known to be damaged—and the limitations of the wellhead equipment, in this case the BOP and components (remember that flex joint).

In my opinion, there was no technical reason to do this static kill. I never heard a rational, cogent argument for what they were getting ready to do, and it was crystal clear that the admiral didn't understand it, or the risks, in the slightest. The only reason I could think of for this procedure was that they were concerned about the worsening leaks on the flex joint and the old BOP and wanted to get the pressure off without flowing the well that would allow for measurement of the volume. Thus, the static kill. It was performed on August 3, pumping 2,300 barrels of mud. It was reported as a success, and BP declared that the well was static, even though the company later reported that they were having to pump 75 barrels of mud every six hours to keep the well "static."[25] During the operation and for several days after, BP disclosed no pressures. Even though the well was clearly not dead, BP, with the agreement of the US government, decided to pump 500 barrels of into it cement on August 5, declaring the job a success. In a briefing on August 6, Suttles reported that they had pumped 300 barrels of cement outside the casing, leaving 200 barrels on the inside, and the operation was "performing as expected."[26]

The media had already checked out when BP announced the "static" condition of the well after the static kill. People paying attention, though, noted that everyone, including Admiral Allen, studiously avoided the terms *dead* or *killed*. They used *static* or *successful* instead. The video feeds grew less informative and less available, as they were pulled offline about this time.

The capping stack and static kill had performed their real jobs: getting the well off national television. They succeeded. The media, with a few exceptions, lost interest. I started getting scolded by strangers on my blog, and some in the media, when I kept asking questions, including asking what the pressure was on the well.

Then on August 10 came the announcement from Admiral Allen. They were going to pressure-test the well, raising the ominous possibility that all might not be right with the world, and that there could be communication from the formation, up the annulus, the space between the production casing and the hole itself, all the way to the wellhead.[27] Clearly, not everyone on the BP–US government team was convinced that the supposedly "static" well actually was. That same day, Wells disclosed that the well, rather than being static, had 4,200 psi on it, but then it got confusing. While Admiral Allen said they were going to pressure-test the well, Wells said they were holding pressure on it, yet it was losing pressure due to "bubbles." The pressure test was actually to bleed pressure off.[28]

For the next several days, they went through what they were calling the "near ambient test," which I never could figure out, except that they were bleeding pressure off the well; then, finally, they undertook the "ambient test," in which they just opened the well up to see what happened. Luckily, the well didn't flow, although it continued to burp oil and gas bubbles, so, on August 16, Admiral Allen said that he had ordered BP to pull the capping stack and the old BOP in order to install a new BOP. His reasoning was that during the eventual bottom kill, they were concerned about pressures exceeding the pressure ratings of three components on the stack. I knew which those were—the flex joint that we've talked about, the riser adapter on top of that, then the transition spool, or connection, between that and the new capping stack—because I, and others in the industry, had been talking about the issue for over a month prior to this. Suddenly everyone was concerned about the integrity of components that had already been subjected to the top kill and the well integrity test and decided to pull the stack and replace it, not really knowing the status of the well or the drill pipe inside it. As part of the procedure, Admiral Allen also announced that BP was going to attempt to fish out the multiple pieces of drill pipe thought to be in the old BOP.[29]

Over the next five days, we were treated to some pretty extraordinary video feeds from both outside and inside the blowout preventer, through downhole cameras. What they found was three pieces of drill pipe and hydrate formation from natural gas coming out of the well and mixing with seawater. Though they were unable to retrieve any of the pipe, Admiral Allen announced on August 27 that they were going to pull the stack anyway and replace it with the BOP from the *Development Driller II*, the rig that had been on standby for months after suspending operations on the second relief well.[30] They got lucky, for once, during this risky operation, successfully pulling the stack and replacing it with a working one (hopefully) on September 4. The old BOP is now in Justice Department control and being examined.[31]

# TEN

# Going for the Kill:
# Firefighting and Relief Wells

In my early days on drilling rigs, I loved everything about it: the challenge, the responsibility, the people, the iron, and the immediate feedback you received, both good and bad. If a job went well, the company man would shake your hand, and sometimes pat you on the back. If it went poorly, well, there was also immediate feedback . . . I'll leave it at that. When I became the company man at the ripe old age of 28, I relished the responsibility and the action. I would stay on location for days, often when we were just turning to the right (drilling). I would definitely be there around the clock if we were having hole trouble, were stuck, were logging, or were running pipe. The hours were often tough, but there was something challenging about even that. My record is six days with no sleep on a stuck pipe job outside of Center, Texas, pumping Black Magic every hour, around the clock. Thank God it pulled loose on that sixth day. I was pretty goofy by then. When I finally got home, I think I slept for 18 hours.

In the years I sat on rigs, I can say that I missed only one logging or pipe job. Again, it was in East Texas, in the winter of 1983, when the roads were so icy that I couldn't even get out of my driveway. The job started about 2 a.m., and I sat by the phone all that night and into the next day, getting a report from the toolpusher every hour, tallying pipe while sitting in my kitchen. I went crazy not being there: first, because it was *my* responsibility, and second, because I loved it. Drilling, logging, running pipe, cementing, fracking, flowing back were *all* exciting, and I hated missing any of it. Hell, I even liked to be there during rig-up and rig-down of drilling rigs, especially the big, high-horsepower rigs. I spent lots of time in East Texas, South Texas, South Louisiana, and New Mexico, as well as inland wetlands and offshore, but I always loved going to other areas where I hadn't worked as much—like Wyoming, Colorado, Alabama, and West Virginia.

Over those years, I came to clearly understand well control and pressure. In order to drill a well, treat it, test it, or produce it, you need a clear picture of what's going on downhole, and you have to understand how pressures you are seeing on the surface tell you what's happening in the well. Through flow rates, pressure on the tubing or drill pipe, and pressure on the backside, or annulus, you can tell what's going on down below if you can visualize it. Well-control school taught kick recognition (to alert you that a well is about to get away from you) and kill calculations, and we practiced circulating out kicks with choke manifolds. In those days, the simulators were pretty primitive, but they gave you an idea of what it was like to experience it for real. Understanding these principles was key to safe operations and staying ahead of the well. Pressure is amazingly destructive. It doesn't take much to kill a man; even a tire rupture while you're airing it up can be deadly, and that's only about 30 psi. In the oil field, pressured vessels are everywhere: wellheads, flow lines, separators, tanks, pumps, and hydraulic systems. Seeing one of those things rupture just once should give a person a healthy respect for the destructive power of pressure and the importance of keeping it under control.

## Oilwell Firefighting: The Masters Emerge

In 1978, the year I moved from working for a pipeline corrosion-control company to drilling rigs, was the same year that Asger "Boots" Hansen and Ed "Coots" Matthews split from famed oilwell firefighter Red Adair to form Boots & Coots, still the best-known well-control company in the industry. Red, whose real name was Paul, was legendary, and a larger-than-life character even though his stature was a diminutive 5 foot 6. Born and raised in Houston, he started in the oil field in 1938, working for Otis Pressure Control, and then was inducted into the army and did bomb disposal. That job obviously gave him a taste for the daring, because upon returning home from the war, Red joined oilwell firefighting pioneer Myron Kinley, who had taught himself firefighting after his father, an old-style California "well shooter," learned that you can put out a fire by exploding dynamite right in the flame. Well shooters were the early well stimulators; by dynamiting the bottom of shallow wells, they would fracture the rock, allowing oil to flow more freely to the wellbore. It was the first primitive frac job, and a dangerous one at that. In 1913, Myron's father, Karl, was on a well that blew out and decided to try putting out the fire with a charge of dynamite. It worked. Setting off dynamite in an oilwell fire accomplishes two things, when done right. First it diverts the flow of fuel, the oil and gas, from the source of ignition; then it momentarily uses up all the oxygen in the area of the fire, snuffing it out. That technique is still used today, but now with an insulated container held over the fire by a long boom, ignited with an electrical detonator.[1, 2]

Back when Red joined M. M. Kinley Company, oilwell firefighting was really dangerous (as if it's not today). Myron himself had been badly injured several times, and his brother was killed in a South Texas blowout in 1938. Red learned the trade from Kinley, soon joined by both Boots & Coots. The oilwell firefighting dynasty had been established. Red worked for Kinley until 1959, when he left to start the Red Adair Company. Boots & Coots both followed shortly after. In 1961, Red and his crew killed a huge blowout dubbed the Devil's Cigarette Lighter in Gassi Touil, Algeria, landing him on the cover of *Life* magazine. While the well was blowing out, John Glenn, who orbited the globe on February 20, 1962, noted that he could see the oilwell fire from space as he flew overhead. Red was so famous in those days that everyone knew who he was; in 1968 John Wayne even made a movie about oilwell firefighting using Red as his role model. Red was an adviser to the movie and provided the firefighting equipment, and his crew even started the fires on the set for Wayne to put out. Wayne and Red became lifelong friends.

Red was well known for his role in killing several other big blowouts, including Phillips Petroleum's (now ConocoPhillips) *Ekofisk Bravo* well in 1977. Fortunately, that well flowed for only eight days from a surface BOP (which had been installed upside down), allowing Red to get to it fairly easily, though flow rates were high. It was estimated that 200,000 barrels flowed into the Norwegian sector of the North Sea before it was capped.[3] Pemex's *Ixtoc 1* well, in the Bay of Campeche off the Mexican coast, blew out in June 1979 and produced up to 30,000 barrels per day into the Gulf. That well blew out for 10 months until it was finally killed in the spring of 1980 with one of two relief wells. During this time, in 1978, Boots & Coots left Red to start their own well-control business, with the obvious (and now very recognizable) name of Boots & Coots.[4]

In July 1988, Red was called to the deadliest oilwell fire in history, the *Piper Alpha* disaster in the North Sea that killed 167 men. *Piper Alpha* was a massive fixed platform 120 miles off the coast of Scotland, in some of the roughest seas in the world. It was operated by Occidental Petroleum (Oxy), based in Los Angeles, run by world-famous oilman Armand Hammer, and was a key facility to their North Sea operations. Producing 300,000 barrels per day at its peak, *Piper Alpha* was converted to gas production in the early 1980s. The platform had 25 producing wells and also served as a central gathering and processing point for gas from the nearby Claymore and Tartan fields. It transported natural gas 128 miles to a processing facility in the Orkney Islands.

The platform consisted of a drilling unit, an oil-production unit, and a new gas-processing and compression unit that had been installed close to the living quarters. The units were separated by firewalls. The platform employed over 200 crewmen at a time, working 12-hour tours. Because of the large processing and compression facilities, the *Piper Alpha* platform was equipped with a massive automatic deluge firefighting system designed to flood the work spaces in the event of a fire. Unfortunately, the system was switched to manual for most of the summer because divers working in the water below could get sucked into the seawater intakes should it engage without warning.

The fire did not start with a blowout or well problem. It started with a natural gas liquids (NGL) pump, one of two on the platform. One had been taken out of service to repair a pressure safety valve, but it had not been returned to operation before the end of the day tour. Rather than complete the work, the crew took the night off, leaving a blind flange where the valve was supposed to be reinstalled. The maintenance supervisor took a report to the control room to inform the production manager that it was to be out of service for the night. The manager was busy, so apparently he just left the report on the control room desk and didn't talk to anyone to make sure they were aware of the out-of-service pump. Naturally, the other NGL pump failed later that night and wouldn't restart. Since the control room staff had not seen the report and didn't know the other pump was out of service, they started it. Pressure immediately built up behind the flange, and it ruptured, filling the work space with vaporized NGL; in moments, another engine or a spark lit it off, and the resulting explosion shut down the platform and closed the safety valves. The problem was, however, that the control room was destroyed in the explosion along with the PA system. There was no communication on the rig, and the off-duty crew in the quarters were unaware of the impending crisis.

As smoke filled the rest of the platform, workers congregated in the quarters waiting for instructions. Access to the lifeboats was cut off by fire, but that was not the real tragedy. The fire would have burned itself out if the *Tartan* and *Claymore* platforms had responded to *Piper Alpha*'s mayday call and shut off their production. They didn't do that. Occidental was highly focused on financial results, and shutting down production was very expensive, because it took several days to get all the facilities lined out after a shutdown; consequences were likely dire for the manager who made a shutdown decision. The organization was so hierarchical that operators on the platforms had to get permission from an onshore control center before they could shut down. Even after two mayday calls, their production managers refused to shut down, even as they saw the explosions on the platform in the distance. Twenty minutes after the initial blast, the *Tartan* pipeline under the *Piper Alpha* platform exploded; an hour later, the *Claymore* line did the same. The two platforms finally got word from shore to shut down, but it was way too late. By the time Red got the call to come put this fire out, most of the platform had melted and fallen into the sea, along with the crew quarters. Ultimately 167 were killed, including two rescue boat crewmen trying to save people in the water. The vast majority of the 59 survivors are the ones who ignored orders and jumped the 100 feet to the water below the platform.[5]

It just so happened that a semi-submersible firefighting platform, the *Tharos*, was alongside the *Piper Alpha*, attempting to fight the fire and rescue the workers. Unfortunately it took 10 minutes to get the water cannons working, and because they were so strong, they raised the fear of injuring or killing crew on the platform. A rescue bridge on the *Tharos* designed to allow escape from the burning platform also failed to deploy, making the rig almost useless in the early hours of the fire. Red later used the rig as a platform to reach the blowout wells and kill them.

One of the strangest twists of this story, though, was the eventual fate of the *Tharos*. It was refitted in recent years as a drilling rig, eventually becoming Transocean's *Marianas*—yes, the very same rig that was drilling the Macondo well when it was damaged by Hurricane Ida in November 2009, and that was replaced by the *Deepwater Horizon*.[6]

The *Piper Alpha* fire was especially challenging since it was very difficult to tell which wells were actually on fire in the twisted wreckage. Red and his crew started to work on the surface, and a young well-control engineer began working on a relief well. This engineer, who ran Eastman Christensen's relief well team, was John Wright. John, a good Aggie engineer from Texas A&M (every oil company needs at least one), had been in the oil field for about nine years when he took this project. A well logging engineer for Schlumberger early in his career, he moved to Eastman Christiansen after a few years and began developing increasingly sophisticated downhole surveying and ranging tools that would greatly improve knowledge of hole location and the success rate on directional wells. He quickly became a relief well specialist, running Christensen's Relief Well Group.

When John arrived on location, a relief well from a semi-submersible had already been spud from a location he didn't consider optimal. Nonetheless, he designed a plan to intercept one of the blowout wells and took over operations of the floating rig. In the meantime, Red was working on the surface; on July 29, he was able to get an inflatable packer into the well and kill it from the top with seawater. John's relief well, which was within two days of intercept, was canceled at a depth of 5,859 feet. Red had beat him to the punch, and finished the job at the surface.

In 1989 John started the John Wright Company, specializing in blowout control engineering, relief well operations, and control of underground blowouts. Rather than firefighting, his focus was on the relief well engineering and planning side, as well as blowout simulation and tool design. In 1991, while he was perfecting well-killing simulations and casing detection tools in Houston, Red, Boots, and Coots, along with competing firefighter Joe Bowden of Wild Well Control, all ran off to Kuwait for the largest well-killing operation in history.[7]

When coalition forces, led by the United States, ran Saddam Hussein out of Kuwait and back to Baghdad, he left scorched earth behind . . . literally. The Iraqi army set fire to over 700 Kuwaiti oil wells, leaving landmines around them to hinder well-control crews as they fled back across the border. Some believe that the whole point of Saddam's invasion in the first place was to punish Kuwait for increasing its production that year, keeping oil prices low.[8] With the Kuwaiti wells flowing uncontrolled, they burned about 6 million barrels per day, fouling the air and polluting the desert floor. Red Adair Company, Boots & Coots, and Wild Well Control were all hired to cap the wells, a job that was estimated to take three to five years. It turned out to go a lot faster than that. Once they figured out the technique, which was mainly deluging each well with water, they just went from one well to the next. The multiyear job actually ended up taking about 10 months in all, at a cost of about $1.5 billion, including the cleanup.[9]

## A Small Band of Experts to Avert Big Disasters

The oilwell firefighting community is a small one, and the relationships among these free spirits are long and complex. Over the years, there have been business arrangements and partnerships, often shuffled multiple times, a tradition that continues even today. I've already mentioned Red splitting off from Myron Kinley in 1959, and then Boots & Coots splitting from Red in 1978. In 1994, Red retired and sold his company to Global Industries of Louisiana; also in 1994, John Wright partnered with Boots & Coots, forming Wright Boots & Coots Engineering. In 1995, Red's senior firefighters didn't like Global Industries, so they split to form International Well Control (IWC), then almost immediately joined forces with Halliburton Energy Services to form WellCall Alliance. In 1996, John Wright split from Boots & Coots, going independent again. In 1997, IWC acquired Boots & Coots, a deal that reunited almost all of Red's senior team from the old days. Months later, they went public and attempted to consolidate the fragmented well-control industry. After hiring an investment banker to the executive team (generally a big mistake, and this time was no different), they took on too much debt, and the company struggled from 1999 to 2005—when Hurricane Katrina basically saved them by wrecking a lot of wells and bringing them a bunch of business in the Gulf of Mexico. Boots &

Coots then expanded into the hydraulic well-control business, making acquisitions in 2006 and 2007. In 2009, they acquired the company of their old friend John Wright, and in early 2010 they announced they were selling to . . . wait for it . . . Halliburton. The circle was complete, at least for a minute.

By now, most of the old legends are gone. Myron Kinley passed years ago, in 1978. Red died in 2004. Joe Bowden died in 2006. Coots died in March 2010. Boots is hanging in there. As with every other industry, though, life goes on.[10–12]

Right after the blowout, Boots & Coots got the call from BP. John Wright was the man they wanted to get a relief well drilled to stop the monster roaring into the Gulf of Mexico 5,000 feet below the water's surface, and he is, by far, the most qualified person on the planet to be running relief well drilling operations. Of the last 83 relief wells drilled, John has managed 40, and all 40 have been successful.[13] He clearly knows what he is doing, and he is famous for being as careful as he is precise. Speaking of John for a newspaper article, Dr. Gene Beck, associate professor of petroleum engineering at Texas A&M, said, "People have total confidence in his ability. I can assure you that from his perspective, every detail will have been well thought out."[14] They clearly have the right guy. At times during the relief well drilling in the Gulf debacle, though, I wondered if he was really in charge.

With a company like BP, as with most huge corporations, you have committees to decide when committees meet. When I was with El Paso, one of the first big mergers we did was with Tenneco Energy, in late 1996. I'll never forget my first meeting with them. It was a planning meeting to work out one of the business unit transition plans. We showed up with four of us, all senior officers, and walked into a conference room that was wall-to-wall people. The conference table was full, as were all of the chairs around the room. There must have been 30 people in there, typical for a big company; the more people you have involved, the more people you can blame. Forms and forms and forms, and lots of lines for people to sign. I'm an entrepreneurial type who likes to move fast, and that merger drove me crazy; so did El Paso as it got larger and larger, evolving into a giant bureaucracy. From what I've seen of BP from the outside, it's very much the same, with diamond-wearing, aristocratic senior executives, and huge fancy buildings with their names plastered all over them.

Relief wells have been used for years to get blowout wells under control, but today they are not really "relief" wells in the sense of relieving pressure in the blowout reservoir. In the early days, that's exactly what they were: When a well was blowing out, another well and was simply drilled vertically as close as possible to the blowout well, then produced at a high enough rate to lower the pressure in the reservoir to stop the uncontrolled flow. Then, in 1933, everything changed; directional relief wells were born, designed to kill the blowout well by applying hydrostatic pressure to get it to stop flowing.

The Conroe field in Montgomery County, Texas, discovered in 1931, was a shallow oil field that produced from sands at around 5,000 feet. The field, ultimately about 19,000 acres in size, rested atop a deeper, heavily faulted salt dome, making the sands diffi-cult to drill and unstable, but with strong deliverability. After years of failed attempts by independent operators to drill the field in the late 1920s, Houston wildcatter George Strake made a deal with Humble Refining (now Exxon Mobil) to drill four wells to 5,500 feet in return for the company bankrolling the project. His first well was successful, coming in at 900 barrels per day, gaining much publicity in the *Conroe Courier*. At this same time, other operators were in the field still, and the unstable nature of the substrata made development risky and well problems numerous. Humble and Strake paused their drilling program after the first well to make sure the state government's field production rules were established to their satisfaction so they could make as much money as possible.

In January 1933, though, they found out just how risky drill-ing in this field was. The previous month, the Harrison and Abercrombie Oil Company (yes, the same Jim Abercrombie who co-founded Cameron Iron Works with Harry Cameron in 1922) drilled the Alexander #1, bringing it in at a respectable 120-barrels-per-day production. But in early January, the casing, wellhead, and tubing all began to sink, rupturing the casing, and the well began to flow uncontrolled. Shortly after, the well bridged over, which is common in wells drilled into unstable strata, and the well dropped to about 10 barrels per day. Harrison and Abercrombie let it go at that. Development in the field continued at a fever pitch, though, with over 60 rigs drilling and 258 producing wells in the field by that time. On June 20, everything in the Conroe field changed when the Alexander #1 unexpectedly roared back to life, swallowing the wellhead, the casing, and even the derrick. The crater forming around the well eventually grew to an estimated 200 feet across and 600 feet deep.

In those days, the state of Texas regulated how much wells could produce, essentially squeezing supply to support higher oil prices. At the time, the total allowable for this field was 32,000 barrels per day. With the Alexander well flowing at an estimated 60,000 barrels per day, the field greatly exceeded its allowable, threatening the revenues for other producers. In the meantime, Harrison and Abercrombie were busily skimming oil out of the crater and selling it as fast as they could. To put a stop to the gusher, and to protect its own interests, Humble partnered with several other operators and bought the gushing well and the 15 acres around it for $300,000. They also agreed to let Harrison and Abercrombie skim as much oil as they could until they got the well stopped.

Humble hired John Eastman, a directional driller from Long Beach, California, to drill a relief well. Eastman, the founder of Eastman Christensen (John Wright's eventual employer in the 1980s), had pioneered slant drilling by the use of a whipstock, which was essentially a metal wedge run into the well that would deflect a drill bit so the well could be drilled at an angle. With this technique, he committed to Humble that he would get his relief well drilled to within a 100-foot circle around the blowout well, starting at a surface location 400 feet away from the gusher. Using tools of his own invention, Eastman first drilled down to 1,960 feet, pulled the drill string, ran the whipstock in to the bottom of the hole, followed by a bit on a deflecting (hinged) tool, and made the turn. After he drilled about 20 feet, they pulled the drill string again and ran in another of Eastman's inventions, a single-shot survey tool that measured the angle of the hole. The process was established. After about every 100 feet of drilled hole, he would survey to make sure his bottom hole location was right and then go back with the bit. Using simple geometry, Eastman knew about where the bottom hole location was. At 4,800 feet, he turned the hole vertical again, drilling into the reservoir. Once to total depth, they set casing, rigged up high-capacity pumps, and flooded the reservoir with water, raising the hydrostatic pressure on the blowout well, and eventually killing it. The first directional relief well was a success, and directional drilling for relief wells was born.[15-17]

## The Macondo's Relief Well

Today the techniques for drilling relief wells are much more sophisticated than in the 1930s and '40s, but the principles for well control and bottom hole location remain the most important. Current survey tools and bottom hole assemblies produce very precise data, giving the directional driller much more chance to hit the target the first time. Modern drilling allows continuous monitoring of the well, what's going on downhole, hole location, formation being drilled, direction, weight on the bit, rotations of the drill string, mud properties, and speed. The driller sits in a literal "driller's chair," controlling all the activities with joysticks and flat panel screens. The assistant driller sits with him in his own driller's chair helping to monitor all of the rig operations.

In a situation like John Wright faces in the Gulf disaster, he has his own screens and data feeds so he can work quietly in an office close by to allow ready access to the driller's shack. These days, most work in deepwater wells is done with rotary steerable tools used to control hole direction from the surface. Once the relief well is drilled down to a certain depth, a turn is initiated with the rotary steerable tool to build angle from vertical. An MWD—measured while drilling—tool is normally in the bottom hole assembly, providing a continuous data feed back to the surface telling the relief well driller exactly where he is. As the well gets close to the blowout well, another tool, run on wireline, uses electromagnetic casing sensing to give a direction and distance to the blowout well. Essentially, the tool, called a magnetometer, is run to the bottom of the hole on electric wireline. It then emits a magnetic field into the rock and onto the pipe of the blowout well. A receiver in the same tool senses the magnetic field emitted by that pipe, giving direction and distance from the relief well bore. This tool is effective up to 30 feet from the blowout well, with very small margins of error. This technique, called ranging, guides the relief well driller to the blowout wellbore with a high probability of success. As the relief well gets closer, shorter sections are drilled before ranging to the blowout wellbore. John Wright describes getting close to the blowout well in terms of golf putting: "You keep getting closer with each shot and eventually you putt it into the hole." Continuing with the golf analogy, and an old saying, he adds, "The more you practice, the luckier you get." [18]

There are two relief wells that have been drilled to be ready to kill the Macondo well, each about half a mile away from the blowout well: Relief Well #1, drilled by Transocean's *Development Driller III* (*DDIII*), was spud on May 2, 2010; and Relief Well #2, drilled by Transocean's GSF *Development Driller II* (*DDII*), was spud on May 16, 2010. Relief Well #2 was drilled to 15,874 feet, with a liner set and cemented, and is standing by in case the first relief well has issues. The first relief well was drilled to 17,909 feet and stood by from mid-July to mid-September despite the fact that it was the only sure kill of the Macondo well. This has been the most confusing decision that BP and the US government have made during this odyssey. The relief well was on track to be easily completed sometime in the last half of July to the first half of August, giving time for weather and mechanical issues, of which it suffered few. With the US government in charge, BP was apparently happy to just sit there with the *Development Driller III*, waiting.

Not until September 13 did BP announce that they were resuming operations on the first relief well—at last. Drilling operations were finally continued after two months of standing by for all of the other operations conjured up by the BP-US government partnership. The well was successfully intercepted on September 17th, confirmed by the relief well losing circulation, which is common, and pressure rising on the Macondo well blowout preventer, indicating communication with the reservoir, which you would expect if there was no cement in the hole. Since there was supposed to be cement in the well after the static kill, it's pretty obvious, then, that that procedure had not been the success it had been declared, and, indeed, the removal of the old blowout preventer was riskier than the unified command was letting on.

The final kill was performed the same day, when cement was pumped into the bottom of the blowout well, and Admiral Allen declared the well officially dead on September 19th, after a successful pressure test. At this writing, the Macondo well, along with the two relief wells, is being prepared for permanent abandonment.

# | ELEVEN |

# The Politics of Offshore Drilling

In the four decades since the Santa Barbara oil spill, offshore drilling has been the bogeyman of the political left, and the much-ballyhooed money machine of the political right. That spill, at 80,000 to 100,000 barrels, tiny by comparison with the *Valdez* and BP spills, shocked Californians and assured a virtually permanent ban on any further development offshore from that state. Ever. The wildlife killed during the Santa Barbara spill was tragic, as in all big spills. And this was one was very public, shown on live television and broadcast coast-to-coast, sparking outrage throughout the country. The outcry was so great that the spill motivated environmentalists to establish the beginnings of the movement to cease all offshore drilling. Despite growing dependence on offshore exploration for our energy supply, the opposition to it has remained vocal for the last 40 years, gaining strength in 1989 after the *Valdez*, and again after this current disaster.

Today the movement against oil and gas development is active, but it is especially vocal when it comes to drilling offshore. Of course, the industry is responding—in fact always responds—that it just "too complicated" for normal people to understand. Rather than have rational discussions, proponents of the industry immediately retreat to their list of pro-drilling talking points, with money always being Point Number 1. The rest of the list is as apocalyptic as possible, including oil shortages, $4 gasoline, and the ever-popular "the terrorists win." The environmentalists then always pull out their list of oiled wildlife, polluted air, and big oil's pillaging of the economy, which is just as apocalyptic, for good reason, but which leaves out mention of the inconvenient question "How do you get to work every day?" Answer: "Oh, I drive my 2009 Cadillac Escalade." The lines are then drawn, both sides ignoring key issues that must be addressed if we're going to solve our own energy crisis.

Like the immigration debate, Americans can't have an objective, unemotional conversation about oil and gas exploration, *especially* when it comes to the offshore. The problem, as with so many other political issues, is that people approach the same subject from very different perspectives. Those who depend on oil and gas for their livelihoods argue vociferously for more and more development.

Obviously, those wage earners dependent on this industry are worried about their income, even though their own lives are threatened not only by oil field accidents, but also by environmental catastrophe in the very places where they live. It could not be more true than in the coastal areas of South Louisiana, Mississippi, Alabama, and Texas. The vast majority of offshore workers live in this part of the country, many of them moving freely from their offshore rig or supply boat jobs to their fishing jobs on their days off. One life threatens the other, and these people are caught in the middle, complaining, on the one hand, about the pollution of their fishing grounds, but protesting a shutdown or limits to the drilling that caused the pollution because the loss of that income is also a threat.

At the same time, money drives the industry and the politicians who are supported by it. Current earnings, contributions, and the stacks of money in their wallets are the prime concerns for both. Since politicians now campaign year-round, they are always looking for ways to stuff their sack full of money by keeping their real constituents, their big contributors, happy. They also take public positions that are popular in their particular jurisdiction but often vote contrary to those positions, worried about slowing down the stream of campaign cash.

## The Political Money Trail

The oil and gas industry hires lobbyists who get paid by stirring up controversies and dragging out legislation by fighting over every provision, no matter how reasonable or justified. The industry encourages them do it, rationalizing their actions by perpetuating the myth that the government never does anything right and that the free market will cure all ills, even when they know that's not true. The industry fights every bit of safety, health, and environmental regulation while scratching their heads and wondering why everyone hates "big oil."

When I was an exploration company CEO, I tried for years to get my industry associations to take more moderate positions on safety, health, and environmental legislation, as well as federal offshore leasing terms. I warned them that fighting all regulatory reform and giving the vast majority of the PAC money to Republicans, who would vote with them anyway, was unwise and shortsighted. On more than one occasion, I was told how stupid I was, that the Democrats hated the industry, and that it was a waste of time even meeting with them, much less giving them contributions. Industry groups also believed the myth that the Republican majority was "permanent." They believed that, of course, right up until the Democrats took the House and Senate in 2006. Suddenly I went from being the crazy uncle the industry kept locked in the attic to someone whose phone started ringing. The problem was, it was too late. Politicians, especially ones who've been stiffed or maligned, have really long memories. In one election, the industry lost its ear in Washington. In 2006, that $120 million the industry had given to the Republicans over 10 years didn't look like such a great investment. The $32 million they had given to the Democrats looked paltry, and it was; the industry suddenly found itself on the outside looking in.

To make matters worse, the 2003 mid-decade redistricting in Texas that was jammed through the state legislature by Tom

DeLay, aided by Governor Rick Perry, Lieutenant Governor David Dewhurst, and then House Speaker Tom Craddick, forced the statewide majority of the US House delegation to suddenly shift from Democratic to Republican by a wide margin. It was part of Karl Rove and Tom DeLay's "permanent majority" strategy, which they had established after George W. Bush's victory in 2000, and some of the gerrymandering was so egregious that one new congressional district ran in a thin line from the heart of Austin over 250 miles to the Mexican border. This brutal redistricting had the specific goals of breaking up Democratic strongholds and marginalizing minority voters. It succeeded, sort of.

Of the 10 Democrats whom DeLay targeted for defeat in 2004, 6 lost their seats and 1 defected. Junior Republicans replaced senior Democrats when the new legislative session opened in 2005, and that was fine and dandy with DeLay and his boys until 2006, when he found himself booted out over ethics and criminal charges, and the Democrats took control of the US House. When the Democrats won control, Texas, the largest oil state, went from having influential Democratic representatives, who were poised to take committee chairs, to junior Republican backbenchers who got the leftovers. DeLay hadn't thought of that happening; he had outsmarted himself, leaving his home state in the lurch and a Democrat, Nick Lampson (one of the politicians DeLay had earlier targeted for termination), sitting in his old seat. So much for "permanent majority."

Traditionally, conservatives have been friends to the industry, and liberals, or progressives, have always been the "opposition." Very few in the industry have been able to establish constructive relationships with the Democrats, though most claim, falsely, that they have. Oil industry lobbyists always boast that they have great relationships and access to both sides of the aisle, but that doesn't often translate to influencing legislation introduced by Democratic leadership.

## Regulation and the Bush Years

Since the Macondo well blowout, there has been a lot of talk in the media and among politicians about the role of the federal government in regulating offshore drilling, especially in the deepwater, and how lax oversight contributed to the disaster. Predictably, right after the blowout, Republicans ran to the nearest Washington television studio to decry the Obama administration's poor management of the Minerals Management Service, demanding that Director Elizabeth Birnbaum, who had been in office less than a year, and Interior Secretary Ken Salazar, who had been in office just over a year, should immediately resign in shame for their horrible mismanagement of this agency that had been run so well for eight years under George W. Bush.

Of course, they also had no tolerance for anyone even mentioning anything that had happened under the past administration and viciously attacked anyone who dared bring it up. In fact, during one of the many congressional hearings about the blowout in May 2010, Doug Lamborn (R-CO) complained that Salazar was "harping" about what the MMS did or didn't do in the previous administration. "Why aren't we talking about the here and now?" he demanded. Salazar, with the perfect comeback, said, "We've done a lot to clean the house at MMS, unlike the prior administration. This is not the candy store for the oil and gas kingdom that you and others were a part of."[1] How true that statement was, but it didn't have much effect. The Republicans continued to bash Salazar and the administration for BP's mess and finally hounded Birnbaum into resigning in May 2010.

To understand the height of the hypocrisy in the Republican politicians' rhetoric, it's worth doing a little "harping" at the past administration to get a sense of what happened to federal regulation of offshore drilling in recent years.

Since both Bush and Cheney hailed from the oil industry, unleashing it from any regulation that impeded its expansion was a top priority of the new administration, and they began to implement it early in the first term. Less than two weeks after taking office in 2001, Bush formed the National Energy Policy Development Group (nicknamed the Energy Task Force), appointing, of course, Dick Cheney as chairman. In the ensuing four months, Cheney and his staff met regularly with industry representatives and lobbyists to help them formulate the task force's recommendations, issuing its report to the president on May 16, 2001. During the task force's work, a huge controversy ensued over Cheney's conduct of the meetings and continued long after the task force had completed its work, as the White House fought to keep secret the names of the participants and companies that met with him and his staff during this time. Ironically, I was one of those people; at least I think I was. I was with El Paso Corporation at the time, and I took part in a meeting in the Executive Office Building, next to the White House, with one of Cheney's staffers. The meeting was about the importance of natural gas supply, as I recall, and my role was brief, the whole thing lasting no more than about 20 minutes; I don't think that anything any of us said in that meeting had any great effect on the task force's ultimate recommendations, and it seemed that we had been given a hearing just to be polite to us. I remember thinking afterward that the whole thing was pretty perfunctory.

On May 18, 2001, two days after Cheney's Energy Task Force report was delivered to the president, Bush issued Executive Order 13212, titled *Actions to Expedite Energy Related Projects*. This order was given to all agencies with this directive:

> For energy-related projects, agencies shall expedite their review of permits or take other actions as neces-
> sary to accelerate the completion of such projects, while maintaining safety, public health, and environmental protections. The agencies shall take such actions to the extent permitted by law and regulation, and where appropriate.[2]

The order also created an interagency task force to "assist the agencies in their efforts to expedite their review of permits or similar actions . . ." As PBS's Bill Moyers pointed out several years later, EO 13212 contained language remarkably similar to that found in an American Gas Association letter to the Energy Task Force delivered in March 2001, which proposed to "streamline regulation of exploration and production on federal lands (including federal waters and the Outer Continental Shelf), while protecting the environment."[3]

In the years after Interior received the executive order, the MMS began waiving environmental review and giving other exceptions to expedite offshore drilling plans. As many as 400 waivers per year were granted to applicants in order to "streamline" offshore permitting as demands for the MMS to move faster grew louder. At the time Bush issued his order, the MMS was already laboring under a congressionally mandated 30-day time limit to review permit applications under the Outer Continental Shelf Lands Act (the law under which offshore permits are issued) amendment passed in 1978. As offshore development stepped farther out into the deepwater, the agency began letting permits through as pressure from industry and the administration mounted.

That kind of shallow review is how you get spill plans for the deepwater Gulf of Mexico, like BP's, that include mitigation plans for seals and walruses—which don't actually live in the Gulf of Mexico. The plans were mostly boilerplate, used by essentially all of the major deepwater players. As deepwater development accelerated, MMS review and oversight decreased. Several times in 2007, the MMS assessed the environmental impact of drilling in the central and western Gulf of Mexico. Each time the agency played down the possibility of a blowout or large spill, estimating in one report that the maximum size of a spill would not exceed 1,500 barrels. In another report it described a large spill as 4,600 barrels, assuring that it would dissipate in less than 10 days and not come ashore.[4]

In May 2010, the *New York Times* reported the growing rift

between the MMS and NOAA scientists during the Bush administration, with scientists accusing the agency of purposefully "understating the likelihood and potential consequences of a major spill in the gulf and understating the frequency of spills that have already occurred there." The *Times* quoted one of the scientists, who wouldn't give his name for fear of reprisal, saying of the MMS,

> "You simply are not allowed to conclude that the drilling will have an impact," said one scientist who has worked for the minerals agency for more than a decade. "If you find the risks of a spill are high or you conclude that a certain species will be affected, your report gets disappeared in a desk drawer and they find another scientist to redo it or they rewrite it for you."[5]

Sound familiar? That's because the same thing was going on at NASA during this time, except that there, Bush's political staffers in the White House were redacting scientists' opinions that supported the problem of global climate change.

Another biologist who left the MMS in 2005 was quoted by the *Times* as saying, "What I observed was M.M.S. was trying to undermine the monitoring and mitigation requirements that would be imposed on the industry."[6]

This controversy between the scientific community and the MMS grew during the Bush years and became even more tense in 2009 when a project supervisor on BP's *Atlantis* production platform, Kenneth Abbot, a contractor, blew the whistle on the company through his lawyer, alleging unsafe operations, subsea production line leaks, and lack of documentation on the platform. He alleged that BP's operation was unsafe and that it had ignored his repeated warnings. To complicate matters, BP laid Abbot off in the fall of 2008, prior to his letter to the company. He has since filed suit against BP for these violations, which, of course, BP denies.[7]

## The Role of Royalties

Lax environmental oversight by the Minerals Management Service was not its only issue during the Bush years. Since its creation by Congress in 1982, the agency has struggled with an organizational conflict that was designed into its multiple missions to oversee oil and gas development on federal lands. The inherent conflict is that, historically, the MMS collects royalty revenues (money) from the very parties it is regulating. The industry, the regulated party, has every incentive, for obvious reasons, to maximize both its drilling and production activities on federal lands while striving to minimize its payments to the government. The Bush administration shared those same motivations with the industry and encouraged the expansion of programs that put the agency at a disadvantage. One was called the RIK or Royalty in Kind program; it was actually started as a pilot program during the Clinton administration.

Before we get into the RIK, a little background. When an oil company leases someone's mineral rights, the most common form of paying that mineral owner is through royalties, or a percentage of the oil and gas sold from the lease. The idea is that the oil company sells its oil and gas and then pays a percentage of those revenues to the owner for its share of sales. This form of payment is fraught with opportunities for the producer to cheat the mineral owner by underreporting production, overreporting shrinkage (loss through processing or fuel use), or overcharging fees to gather and process the gas. Many sophisticated mineral owners build safeguards into lease agreements to protect against oil company abuse, and some actually do take royalty in kind.

Royalty payments on federal lands work in the same way. The government owns the minerals (oil, gas, and coal) beneath federal lands and collects royalties for their sale. Over the years, there were many disputes about payment of royalties and the value that the government was receiving. To help resolve these disputes, a pilot program was devised in 1997 in which the

MMS took production *in kind*, or essentially got paid by taking the oil and gas and selling it themselves. The problem? Nobody at the MMS knew how to do that, and they didn't hire qualified people to manage it. The oil and gas trading markets are very sophisticated, and novices will get fleeced quickly. You could classify employees at the MMS as novices when it comes to oil and gas trading, and large energy companies often dominated the markets they sold into with armies of traders.

As part of Bush's new energy policy, implemented by Cheney, the pilot program was rapidly expanded, exposing as much as half of all royalties received by the MMS in 2008 to this trading risk.[8] Almost every year from 2003 to 2009, the GAO, or General Accounting Office, pointed out weaknesses in the RIK program, citing lack of proper training, inadequate reporting and auditing standards, overwhelmed and discrete IT systems to track production and royalties due, and the weakness of allowing companies to self-report, along with their ability to change their reports for up to six years after production without prior approval from the MMS.[9] Secretary Salazar announced an end to the RIK program in September 2009, winding down the remaining contracts over the next year. When he made the announcement that he was shutting the program down, Salazar noted, "The royalty-in-kind program in my view has been a blemish on the department."[10]

## Revolving Doors and Cozy Relationships

During the Bush years, government agencies, especially the Department of the Interior, became huge, rapidly revolving doors between industry, lobbyists, and government offices. Lobbyist involvement in writing legislation, guiding policy, and buying influence was raised to an art form during these years, driven by characters like Tom DeLay in the House of Representatives, the notorious Jack Abramoff on the outside, and the Bush crew in the White House. One notable case of this abuse was Deputy Director Steven Griles, a coal industry lobbyist who was appointed in 2001 under Interior Secretary Gale Norton; Griles ended up being Interior's senior representative to Cheney's Energy Task Force. When he was approved for his appointment, Griles was allowed to go into government service under an exception with the Office of Government Ethics and the US Senate that allowed him to receive both his government salary and over $1 million in payments from his former lobbyist employer, signing an agreement that he would recuse himself from "any particular matter involving specific parties in which any of [his] former clients is or represents a party." Despite this agreement, it was revealed in 2002 that he had indeed continued to meet with industry representatives and that he was tied closely with Abramoff. At the end of 2004, during the investigations into his conduct and relationship with Abramoff, Griles resigned from Interior. In 2007, he pled guilty to one count of lying to the US Senate and was sentenced to 10 months in prison and a $30,000 fine.[11] Of course, we all know that Abramoff served 43 months of a 70-month sentence after pleading guilty to charges of tax evasion, fraud, and conspiracy. He was also fined $21.7 million for restitution and cooperated with investigators.[12]

In May 2010, Interior's inspector general issued a report of their investigation into activities within the MMS during the Bush years that included details of MMS employees accepting tickets to sports events, gifts, and trips offered by oil and gas companies, all in violation of federal laws. This report followed a scathing one issued two years previously by the inspector general that documented dozens of MMS employees accepting invitations for travel, golf tournaments, and other events including parties thrown by industry representatives; detailed in the report were drug use, sex, and I can only assume a little rock and roll. The government employees participated in outings so frequently that industry representatives labeled them the "MMS chicks." It seems as though the "expedited review" ordered by President Bush, managed by his political appointees, created a culture of *party till the sun comes up,* at least for some in the MMS.[13]

## New Reforms Under Way

Also in May 2010, Salazar proposed new safety guidelines for approving and operating offshore wells. One of his key proposed changes was to extend the 30-day statutory review limit to 90 days. With 20/20 hindsight, this is one proposed change that will likely be approved in Congress, and it is wise. It will provide some statutory cover and give the new BOEMRE time to actually review APDs (applications for permit to drill) rather than just rubber-stamping them and putting them in the file. Additional safety improvements that Salazar proposed are enhanced deepwater well-control procedures—which is a lot easier to say than actually do—new casing design and cementing procedures, and development of more effective response to deepwater blowouts. He also addressed new blowout preventer systems and procedures, as well as increased enforcement of existing rules and regulations.[14] All of these changes and improvements will take time—hence the six-month deepwater moratorium that he has proposed (which is really not long enough to enact all the changes and redesign). I don't expect that to stick, though, now that BP has killed the Macondo well and all is right with the world as far as the industry is concerned. There are already noises coming out of Interior that the deepwater moratorium would not likely extend past the November 30 deadline, and could be lifted earlier depending on industry cooperation with BOEMRE's new workplace safety, well control, and spill response guidelines. In a trip to Biloxi, Mississippi, in early September, new BOEMRE director Michael Bromwich told a town hall meeting of local officials and industry officials that the timing depended on the industry complying with current and new regulations.[15]

At this writing, the CLEAR Act (Consolidated Land, Energy, and Aquatic Resources Act of 2010) was approved in the House on July 30 and is currently languishing in the Senate. The most important provisions of the act were the lifting of the $75 million liability cap, redundancy requirements for blowout preventers, and independent certification of new casing design requirements. It's having a rocky road in the Senate, but I expect some form of the bill to eventually pass, along with a raised liability cap. I don't expect the liability to be unlimited, but it will likely be raised enough to give small independent offshore operators pause about whether they want to continue in the Gulf of Mexico. These new rules and regulations will likely drive a wave of consolidation in this segment of the oil and gas industry, where large operators like Shell, Chevron, and even Anadarko could conceivably gobble up the smaller players. The offshore business will either go through this consolidation phase or develop some kind of risk sharing, pooling interest to spread risk over more companies.[16, 17]

One thing is fairly certain, though. We won't go back to business as usual in the industry or within government agencies. This blowout and environmental catastrophe have shaken industry, government, and the American people. We all need and want abundant energy, but at what price to our futures? One thing that we learned (or should have learned) is that if companies are to drill in these challenging environments, we must demand strict guidelines and procedures, comprehensive spill response, and strong government oversight to minimize the possibility of this happening again. It's the one lesson that we should have taken after the eight years of lackadaisical oversight from the previous administration. I am concerned, though, that all the happy talk of "disappearing oil" coming out of the administration will weaken the resolve of Congress and regulators to actually fix the problems that we face in the offshore.

The corruption and negligence in the MMS during the Bush years is well documented. As Matthew Yglesias put it in May

2010, "This, of course, was the very essence of the Bush admin-istration approach to government. When a regulator could be staffed by shills for the industry it was supposed to oversee, it was. When no industry particularly wanted to own an agency, like FEMA, it was handed over to a random crony. The results were disastrous and we're still paying the price today." [18] Why, then, is the Obama administration taking all the blame? That's an easy answer: politics. The Republicans and past Bush admin-istration officials are shoveling as fast as they can to heap the blame on the Obama administration for poor management over the previous decade at the MMS—and many other agencies, for that matter—to distract a public that is not paying attention to what's really going on.

That said, the current administration is not completely without blame here. Even though BP's Macondo well application was given categorical exclusion from the National Environmental Policy Act's requirement for a detailed environment impact assessment before Birnbaum took office, the well began operations while she was director, and no review, to my knowledge, was ordered by her. However, in her defense, most of the people hired during the Bush years were still in place at the agency when she took office, though personnel problems were becoming well known in the agency during 2009. I don't believe Salazar or Birnbaum made changes to the agency as quickly as they should have, but the culture of deference to industry (where all these people wanted to work) as well as poor data collection and oversight had been deeply ingrained within the bureaucracy during the Bush years. The Bush directive to expedite drilling permit processing, on top of the existing 30-day statutory limit, severely limited the agency's ability to review anything in detail, especially since over half of the recent APDs were for deepwater wells.

The Bush legacy of the rubber stamp for permits at the MMS was probably not directly responsible for the Macondo well blowout, but it certainly didn't help, and had there been more thorough review, more blowout preventer inspections and certifications, and closer oversight of well design, this may very well have been caught before it happened. In May 2010, Congressman Ed Markey put it quite well when he talked to the *Washington Post* about lax oversight at the MMS:

> I'm of the opinion that boosterism breeds complacency and complacency breeds disaster. That, in my opinion, is what happened.[19]

How true.

## | TWELVE |

# The Aftermath:
# What Do We Do Now?

No one (except for industry apologists) will dispute that the disaster on the *Horizon* caused the largest environmental catastrophe in the history of the United States. The massive flow of millions of barrels of oil and billions of cubic feet of gas into an open ocean environment has caused untold damage that will take years, if not decades, from which to recover. It will now never be fully understood or measured.

We were witnessing 24/7 during the spring and summer of 2010 the fruits of four decades of neglect and cowardice by our elected leaders, who have been unwilling to risk their own reelection by actually doing their jobs and making some hard decisions. They have instead used energy policy as a weapon against their political opponents, all calling one another liars, pretending to occupy some moral high ground that doesn't exist, and perpetually putting off action that would benefit the nation until after the next election. It's a vicious, never-ending cycle.

This disaster has captured the nation's attention about energy once again. Hopefully, this time it will help change the conversation about how we power our nation's future and what our children's and our children's children's energy future looks like. Right now, our future looks grim, and unless this disaster changes the conversation, I don't see it getting any better in my lifetime. However, in order to fully understand the gravity and scale of this catastrophe and how it will affect our future, it's important to review where we've been since April 20—the good, the bad, and the ugly. This one is mostly ugly.

Off the top, and a fact that we should never forget, we lost 11 dedicated, hardworking men that night on the *Deepwater Horizon*. From everything we've reviewed and now understand, those deaths were the result of complacency, hubris, overconfidence, and bad judgment, qualities that are not conducive to safe operations in one of the most challenging drilling environments in the world. We also know that the abilities of the crew of the *Deepwater Horizon* and its safety systems were woefully inadequate to handle a massive influx of gas. Transocean demonstrated their own lack of urgency and their complacency bred of a familiarity with their own operation. Bypassed or inhibited safety systems that placed an emphasis on the human–machine interface created a critical failure that likely cost the workers' lives, and indeed the entire rig. Had the emergency shutdown system been functioning on the deck engines, those engines probably would not have oversped, resulting in at least one ultimately exploding. Had those engines simply shut down when gas entered their air inductions, perhaps the crew would have had a chance to at least get off the rig floor and out of the pump and mud rooms, if not actually given time to get the well under control before it exploded. Had the general alarm not been "inhibited," perhaps an orderly evacuation of the rig may have been at least begun and serious casualties avoided. Besides the rig floor and mud room deaths, the two worst injuries occurred in the crew quarters areas. Buddy Trahan, one of the visiting Transocean executives, as well as Wyman Wheeler, an off-duty toolpusher, both suffered broken bones and burns from the explosions. Had they been given a chance to evacuate with a gas alarm, perhaps they would not have been injured. Had the crew quarters not been located where they were, or the walls more fortified, perhaps the injuries would not have been so great.

With all the bad circumstances of this incident, there were several mitigating factors that saved many lives that night, none of which was due to good planning. First, the *Damon Bankston*, equipped with a fast rescue boat, was on scene. A quick-acting crew pulled many out of the water as they jumped from the rig floor. The other factors were wind and weather. The seas that night were calm and relatively warm. Had this disaster occurred in January, during a cold front that was whipping up 5- or 6-foot seas, fatalities could have easily reached the dozens.

## What Really Went Wrong?

Many factors have been cited as culprits that caused this blow-out. I personally believe the cause was primarily human error; the managers on the rig simply failed to listen to the well as it became more and more dangerous. In the oil and gas business, many—including me—believe that wells can talk; you simply have to understand the language they're speaking and what they are saying. They speak in many ways; they can tell you what's going on down below in the way they circulate, in the way they build and lose pressure, in the way they flow (or don't). If everything is stable and the well is happy, circulation rates are constant. If there is gas in the well, the background level is controlled and remains steady as it comes up from the bottom entrained in the mud. If the well is unhappy, it will kick into the wellbore a bubble of gas that expands as it comes to the surface; you have to recognize the kick and be ready for it to control the flow as it is circulated out. Sometimes the well will take mud, which lowers the hydrostatic pressure on the formation below, allowing a gas bubble to come in; when that gas comes to surface, it will then kick again.

A really strong well, like the Macondo well, can be a bear to drill and control. It will kick and then lose circulation if you heavy up the mud too much, and then kick you again. If you're not paying attention or are impatient, the well will get away from you when this happens. In drilling big, deep wells like these, inattention, rushing, and impatience have no place. Shortcut a procedure or ignore a warning sign, and the well will come see you. The same is true of wells both onshore and offshore. The problem with the deepwater offshore, though, is that you have a mile or two of pipe on top of the blowout preventer, the control point, which complicates well-control procedures. In 1997, Larry Flak of Boots & Coots wrote an article in *Offshore Magazine* about kick prevention and well control in the deepwater; he concluded, "Blowout control options in ultra-deepwater are very limited. Blowout prevention is of paramount importance."[1] In other words, the best way to control a blowout in the deepwater is not to have one.

There's no question that the Macondo well was managed badly in those final hours. But compounding this, BP drilling engineers made a significant mistake—probably their biggest—in their choice of casing design. By using a long string design, rather than setting a liner, BP did not install a downhole barrier, except the cement, which could prevent gas from coming all the way up to the surface on the *outside* of the casing rupturing the seal assembly. A liner is essentially a length of casing that covers only the amount of open hole in the bottom of a well. At the top of that liner, which comes a short way up inside the last string of pipe, is a device called a liner hanger, which contains a seal. After the casing is cemented on the outside, the liner hanger is set and then sealed. Another length of pipe is run down to the top of the liner, cemented, and locked into the downhole hanger. In addition to the downhole barrier, another is placed at the top of that last length of the tieback casing at the surface. What this design does is provide three barriers: the cement at the bottom covering the productive formation; the liner hanger downhole, right above the producing formation; and finally the casing hanger at the surface as insurance. Even if the cement job fails, the two barriers prevent gas from reaching the surface. In a presentation made to the Aspen Institute's Aspen Ideas Festival in July 2010, Joe Leimkuhler, a senior engineer for Shell Americas, explained the differences between BP's casing design and Shell's.[2] The primary difference was that Shell used the liner, creating that second barrier. Had BP done the same, this disaster might not have occurred.

BP, not surprisingly, disagrees. In an internal report released shortly before this writing, BP drew contrary conclusions that, I'm sure by sheer coincidence, spread the blame for the blowout around to as many parties as possible. The report, while containing much data, attempted to use that data to cast BP in the best possible light. Although the firm took some blame for incorrectly accepting the faulty negative pressure test results and not using the recommended number of centralizers, in this report they also staunchly defended the long string casing design, calling it common industry practice that did not contribute to the initial blowout—though they acknowledged there could have been some flow up the backside after the initial kick. I personally find the logic they used to draw that conclusion pretty tortured, intended to put the majority of the blame squarely on Transocean, Halliburton, and Weatherford (which had supplied the cementing equipment), deflecting fault from themselves.

The report listed eight primary findings that, but for two, placed total blame on other parties. Here is a brief summary:

1. The annulus did not isolate the hydrocarbons— Halliburton's fault, though BP should have been more aware.
2. The shoe track (bottom section of casing) did not isolate hydrocarbons—Halliburton's and Weatherford's fault.
3. The negative pressure test was accepted although well integrity had not been established—primarily Transocean's toolpusher's fault; BP's fault that a company man accepted the toolpusher's explanation.
4. The influx of hydrocarbons was not recognized until hydrocarbons were in the riser—Transocean's fault.

5. Well-control response actions failed to regain control of the well—Transocean's fault.
6. Diversion to the mud gas separator (rather than directly overboard) resulted in gas venting onto the rig—Transocean's fault.
7. The fire and gas system did not prevent hydrocarbon ignition—Transocean's fault.
8. The blowout preventer's emergency mode did not seal the well—Transocean's fault.

Of these eight primary conclusions, BP only takes partial responsibility for two and, not surprisingly, blames the dead toolpusher and contractors for the rest. Conspicuously missing from this list are the casing design; the decision not to run all the centralizers; and the decision not to circulate the well from the bottom before cementing, which could have improved the cement job. Also not on the list was BP's orders to displace the drilling mud in the riser with seawater prior to completing all operations. The report staunchly defended the casing design and concluded that BP well site leaders' decision not to run the specified number of centralizers did not contribute to the blowout, since they contend the blowout initially came through the Halliburton cement and Weatherford cement equipment rather than up the annulus.[3] In my view, passing the buck so obviously compromises the validity of the entire report and its conclusions, though I agree with some. The casing design clearly increased the risk of blowout by excluding a downhole barrier, though I don't believe the centralizers played as big a role as cement volume and composition did. Certainly, a non-nitrified cement would have done a better job of isolation, but again, that was BP's decision. The decision to not circulate bottoms up was also BP's decision. The key mistake here, however, was failure to recognize the influx of hydrocarbons into the well. BP can blame Transocean and the others all it wants, but it was that mistake that led to the disaster, and that responsibility lay on BP's shoulders as operator.

## Assessing the Damage

Let's revisit some key facts after the blowout. On July 15, 2010, BP shut in the well, risking further mechanical damage, but fulfilling their primary goal—avoiding empirical measurement of the total production from the well. Enabled by the US government, BP successfully avoided measuring how much oil was spilled. That means that BP is now in a position to argue that the flow from the well was much lower than what the experts estimated it to be—and thus sidestep liability for some of the damage they caused.[4] Since EPA fines and other liability are based on barrels of oil introduced into the water, it was critical for the government to have an accurate measure of that flow. The containment facilities BP was ordered to install would have done just that, and I believe that BP was able to avoid that obligation by slow-playing construction, as well as delaying installation of the control device called the capping stack. This stack sat on the dock onshore for weeks before BP was ordered to install it. Only after the government compelled BP to move ahead with the new cap on July 9 did the company and the government announce a last-minute "well integrity test" on July 12, 2010.[5] This new test, originally intended to last 6 to 48 hours while final installation of the production facilities was completed, was in actuality the permanent shut-in of the well. BP never completed the containment facilities and, with the government's consent, kept the well shut in, even as the wellhead itself leaked. Even though pressure leveled off at around 6,900 psi after two weeks, we never really knew whether it didn't go higher due to depletion in the reservoir, well damage, or leakage elsewhere. And we never will know, since they came up with another last-minute procedure on July 19, 2010, dubbed the static kill—essentially another top kill procedure, but utilizing the closed capping stack.[6]

The ecological disaster created by this blowout and the subsequent spill is the long-lasting legacy of this story. For the first time ever, massive quantities of the dispersants Corexit 9527 and

Corexit 9500, both banned in the UK, were applied by aircraft, by boats, and, worse, at the seafloor. Even after orders from the EPA to restrict the use of these dispersants, BP continued, proclaiming they had no other choice. The EPA capitulated., and allowed the massive pollution to be made even worse, using 1.5 million gallons, with 800,000 gallons of the chemical sprayed directly into the stream at the wellhead. As we know, dispersants are used to break up oil into tiny droplets so it can be biodegraded; however, this chemical was designed for use on the surface to disperse spills from fuel oil and other petroleum products spilled from ships. In this case, since the dispersed oil doesn't come to the surface, biodegradation can be much slower, causing the droplets to become entrained in the water column, eventually clinging to plankton and other micro-marine life and entering the food chain. We still don't know the long-lasting effects of this decision, and we won't know for years.[7]

Once the well was finally shut in, the oil on the surface of the Gulf that wasn't boomed up rapidly began to dissipate. The media declared that the oil had magically disappeared, citing so-called experts who claimed that it biodegraded much faster than anticipated. They said the water was much warmer than we experienced in, say, the *Exxon Valdez* spill, and that there were more active or even new bacteria. It's not true, of course, that the oil is gone and the destruction is over, and most serious scientists are very concerned that the unseen, untracked, and unmeasured dispersed oil has already entered the food chain, with catastrophic results. They also have said that oil plus dispersant is more toxic and more damaging than oil left alone to simply rise to the surface and be skimmed or biodegrade on its own.

We do know that the government and BP made the determination early on to sacrifice the deep-ocean water column that can't be seen in order to save the marshes, which can. Dispersed oil will stay below the surface, invisible but just as destructive as it enters the water column. We will likely never know the true extent of the damage to the Gulf, especially since the media has lost interest. We will only be able to indirectly see the damage to our Gulf—by watching fishermen, oystermen, and shrimpers go out of business due to low catch records, or by experiencing the rising price of seafood in our supermarkets and restaurants since more will have to be imported from other regions. We'll also see it on our beaches as dead sea life washes up on shore. Or it may be visible in what we won't see, like Ridley sea turtles not coming up on our beaches to nest.

The fishing industry in the Gulf of Mexico is one of those businesses that most people don't think about even as they dine on oysters, shrimp, snapper, and any number of commercially caught fish. It's one of the richest fishing regions around the United States; some would say (at least before this oil spill) that the coastal Gulf of Mexico is among the most productive in the world. Many years and hundreds of millions of dollars have been spent trying to restore natural habitat that had been destroyed by dredging and commercial development of Gulf coastline. However, even as much of the coastline continues to subside, or sink into the Gulf due to dredging and the actions of wind and waves on the marshlands, the water has remained rich with sea life. Now, though, many of those regions are unsafe to fish, and when you travel along the coast, especially in and around the Mississippi Delta, birds, dolphins, and fish are few and far between. Indeed, when I traveled out South Pass at the very mouth of the Mississippi River in June 2010, there were few birds, no dolphins, and no fish visible. I just hope they were smart enough to move to safer areas—if such a move was even possible—rather than staying there to die.

## Displaced Fishermen, Strained Tourism

It seemed for a time that the entire Gulf Coast worked for BP as the Vessels of Opportunity program was implemented, employing fishermen to use their own boats, replacing their shrimp and fishing nets with booms, assigned to skim oil out of the shallow-water areas. Thousands of these fishermen reluctantly joined the BP workforce, just trying to make ends meet as more and more areas were closed to fishing of any kind. They hated it, but did it to save their boats and their businesses.

Tragedies bring out both the best and worst in people. This tragedy was no different. Even as commercial fishermen whose families had worked these waters for generations scrambled for the BP jobs, they also complained of people with brand-new boats suddenly showing up, claiming to be displaced fishermen so they could get BP to pay them for their new vessels. To BP, a fisherman's a fisherman, and they gave some of these newcomers cleanup jobs. Tensions were so high in places like Port Fourchon and Venice that police brought in reinforcements just to keep the peace. Fights among workers were so commonplace that portable jails, constructed from shipping containers, were brought in on flatbed trucks and set in strategic locations just to have someplace to put people to cool off. When I was in Venice in June, a fight broke out between workers in the parking lot right in front of my hotel, normally a peaceful spot for fishermen and their guides to meet and drink beer.

Just like the fishing industry, tourism suffered greatly from the oil spill and will continue to suffer for some years to come. In the late spring and summer of 2010, workers in white hazmat suits shoveled tar balls and emulsified oil, often outnumbering the few tourists who sunned themselves on the beaches. Indeed, tourists were so scarce in South Louisiana that the hotels and restaurants would have been closed were it not for the tens of thousands of cleanup workers who brought their own mini-boom to resort towns, needing places to stay and eat. And drink. Sportfishing was shut down during this time, and in places like Venice, Louisiana, guides just sat on the decks of their tied-up boats taking cancellations by e-mail and worrying how they were going to make their note payment. Along the sugar-sand beaches of Pensacola and other Florida resorts, hotel operators resorted to webcams and television, desperately trying to get people to "come on down" with assurances that there was no oil on their beaches (at least at the time).

It was all in vain. The summer 2010 tourist season for the entire Gulf Coast, from Florida to even some parts of Texas, was a bust. Hotels stood virtually empty for Memorial Day and the Fourth of July. The only businesses that hung in there were bars where locals and cleanup workers went to drown their sorrows. On the heels of devastating hurricanes, especially Katrina in 2005, people in these regions were just getting back on their feet. BP knocked them right off again, and they continue to struggle at this writing.

## Government and Industry: Where to Next?

The federal government has been rocked to the core by the Macondo well blowout. Only weeks before the disaster, President Obama had announced the lifting of a 40-year moratorium on drilling offshore along the East Coast, in the eastern Gulf of Mexico, and in areas off Alaska. Then came the blowout and the largest environmental catastrophe in US history. Thanks a lot. Scrambling to respond, Secretary of the Interior Ken Salazar, who had already been grappling with a scandalous report about misconduct within the Minerals Management Service during the Bush administration, announced in late May that he was splitting the agency into three distinct functions:

1. The Bureau of Ocean Energy Management, to handle planning, permitting, and leasing.
2. The Bureau of Safety and Environmental Enforcement, responsible for oversight of safety and environmental programs in offshore operations.
3. The Office of Natural Resources Revenue, which collects royalties from producers and conduct audits.[8]

Salazar's goal in this hastily designed structure was to eliminate conflicting functions within the agency. I can understand this desire, but it is concerning that we have just created two new bureaucracies rather than fixing the one we had. The main goal here was to improve safety and enforcement of regulations. Simply moving the safety compliance function to another agency like OSHA or EPA would seem to me a much more streamlined solution than creating all these new alphabet agencies, but there I go again, being rational.

The oil and gas industry, at least offshore, has also borne the brunt of BP's disaster; after the blowout, a skittish Interior Department shut down all offshore drilling permits as well as ordering all deepwater rigs to suspend operations. Even though it was later decided that shallow-water drilling could recommence, the new Bureau of Ocean Energy Management, Regulation and Enforcement has been slow to get to its feet and establish new drilling criteria. It is seemingly paralyzed; even though Interior continues to say that there is no moratorium on shallow-water offshore drilling, BOEMRE, at this writing, has yet to issue a single permit for a new well.

In the meantime, the moratorium for deepwater drilling has staggered forward, challenged in the federal courts and the court of public opinion. I support the moratorium, but I believe it's not long enough to really fix the technical issues with deepwater drilling. However, neither the government nor industry is making the best of this pause in drilling. The president appointed a commission to study the blowout and its causes, but this has so far focused primarily on the spill and its effect on the local economies of the Gulf Coast states as well as the environment. That's not surprising, since no one appointed to the commission actually knows anything about what they are supposedly investigating. Not one member has any drilling or production experience, much less familiarity with operations in the deepwater. In fact, the only person on the seven-member commission who has an even passing familiarity with the industry is former EPA administrator Bill Reilly, who happens to sit on the board of directors of ConocoPhillips. I don't expect the presidential spill commission—made up of environmentalists, academics, and politicians—to come up with any material conclusions or recommendations beyond their already well-known preconceived notions.

Unfortunately, the president and Carol Browner, the director of the White House Office of Energy Policy and Climate Change, both missed an opportunity to communicate Obama's commitment to seriously developing a comprehensive energy policy and defining how deepwater development fits into that strategy. No matter how many hearings you have, you're not going to determine the cause of a blowout in 5,000 feet of water by interviewing politicians and the head of the Louisiana Shrimp Association, as important as their issues may be.

At the same time, the industry—instead of doing something constructive—brought out its army of lobbyists to complain about the Obama administration and the unfairness of their treatment. The industry groups also got into the act. Jack Gerard, CEO of the American Petroleum Institute (and a professional lobbyist), put out a long statement in opposition to the moratorium, saying in part, "Freezing access to an important piece of our nation's energy supply impacts jobs and the economy in the Gulf Coast, which has already suffered from the spill, and threatens the nation's energy security without raising or improving industry procedures."[9] Everyone I talked to in the industry also complained about the moratorium, but not one single person could answer this question: "What problems on the *Deepwater Horizon* caused every single safety system to fail, and how do you fix them?" Of course, no one yet knows the complete answer to that question, but the industry remains focused on earnings forecasts and money in the till right now.

For messaging, industry lobbyists and CEOs pulled out the usual weapon of choice: employment. Drilling contractors threatened to pull their floating rigs out of the Gulf, firing thousands. Service companies threatened the same. Bobby Jindal, governor of Louisiana, even held a political rally (the unofficial kickoff, it seemed, for his 2012 presidential campaign) using worried oil field workers as props.

The problem of employment during the moratorium is certainly real; however, it is made worse by the very companies that are complaining about it. For example, drilling companies almost always flag their rigs in foreign countries, such as the Marshall Islands or Panama. The Marshall Islands lists 2,200 vessels under its registry; of those, 117 are MODUs (mobile offshore drilling units).[10, 11] The reasons? There are many—including safety inspection requirements; the ability to contract for a facility inspector of choice; lax wage, employment, and staffing requirements; and avoidance of US taxes. If these rigs were flagged in the United States, these requirements would be more stringent—but most important, they would be required to have a crew made up of at least 75 percent Americans. If the rig moved to foreign waters, the jobs would go with it. As it is now, the drilling company can fire the American crew, hire cheaper labor where they go, and then complain publicly about how the US government caused lost jobs.

To formulate an industry response to the disaster, the American Petroleum Institute did organize a confusing plethora of four industry task forces in May 2010, with overlapping directives to come up with new techniques to drill the offshore. The multiple task forces were obviously to make it look like the industry was all over the issue. However, several important sets of proposals did come from this flurry of meetings, primarily in the area of well containment. A consortium made up of Chevron, Exxon Mobil, ConocoPhillips, and Shell (not BP) announced in July 2010 a commitment of a combined $1 billion investment in the Marine Well Containment Company, which will operate a fast-deployment subsea system designed to contain deepwater spills.[12] Many of the lessons learned from this spill will be implemented in this new system, which will be designed to totally contain out-of-control wells releasing oil into an open-ocean environment. In September 2010, the Subsea task force issued its report, which included not only this proposal but also a protocol for LMRP redesign as well as an industry stockpile of BOP components and connections that will be on hand for rapid deployment when a well gets out of control. The second set of proposals, from the Oil Spill Response task force, illustrates just how difficult it is to recover oil spilled into the sea. The proposals called for more training of personnel and coordination with the government but defended both the use of dispersants and the in situ burning of oil on the surface.[13] I just hope these proposals work better than the last oil industry joint project—the Marine Spill Response Corporation formed after the *Exxon Valdez* spill. Supposedly, the MSRC's $80-million-per-year budget would give it the capability to clean up a massive offshore spill such as BP's.[14] It became painfully obvious very early on that it couldn't manage even a small percentage of its touted capacity.

The key lesson that should be learned from this disaster is that government policy designed to improve our own energy security is badly needed and long overdue. The painful lack of leadership from both parties and presidents over four decades has driven the United States into dangerous dependency on foreign sources of oil that put our security and safety at great risk. Uncontrolled growth in demand, policies that encourage squandering our resources, and lack of courage to stand up to the special interests for the long-term good has put us in a position of weakness, forcing us to be reactive rather than proactive in response to world-shaking events. Our government policies, as well as those of other Western countries, continue to keep the Middle East unstable while fueling terrorist organizations that prosper in some of the very countries from which we buy oil.

Even today we allow our energy security to remain privatized and outsourced. Our politicians talk about moving to renewable sources, conservation, improved efficiencies, and mass transit—all green job buzzwords—but they don't really do anything. The Republicans label every initiative for clean energy or reducing emissions as some form of tax, their favorite bogeyman, and all efforts to extend the decline curve of our conventional fossil fuel resources are labeled by the Democrats as "big oil" raping the land and poisoning the populace. In the meantime, we do nothing but yell at one another on cable news television.

We certainly must reduce our dependency on fossil fuels, especially oil. It should be painfully clear to us all that the extreme environments where we must now go to get more oil present a clear threat of ecological catastrophe if something goes wrong. I don't expect the oil industry to ever move beyond its own demands for quarterly performance and its millions of dollars spent on lobbyists and campaign contributions, tainting our elected officials and thwarting reform. Only if we elect leaders who will take the necessary steps toward reform will we ever have a chance of moving beyond the precariously fragile state of dependency in which we now find ourselves.

The answer here is comprehensive, and not easy. Our energy security certainly lies in not only moving away from fossil fuels, but also moving toward alternative sources of energy, *including* nuclear power, if we can make it safer through tighter controls and standardized, next-generation reactors. Certainly, if France can do it, so can we. In the meantime, we need actual leadership from our leaders to help put in place economic and tax incentives for wind, solar, and renewable fuel sources, as well as immediate conservation efforts such as strong fuel efficiency standards and huge infrastructure projects for mass transit and green buildings. Rebuilding American manufacturing for construction of solar panels, wind turbines, batteries, electric cars, natural-gas-powered trucks, and energy-saving construction materials would not only create millions of jobs but also reignite our economy, accelerating growth and improving our balance of trade and reducing debt. All of this is possible if we demand it of our elected officials rather than tolerating their perpetual two-year cycle of rhetoric and reelection.

I am hopeful that these are the lessons that we learn from the tragic events surrounding the disaster on the *Horizon*. This catastrophe should be a wake-up call to all of us about the importance of our stewardship of the environment in which we live, the importance of the health of the giant ecosystems that feed us, and our duty to never again threaten those vital resources by polluting them with oil or the purposeful introduction of massive quantities of toxic chemicals to hide that pollution from view. We, as a society, must be motivated to move beyond our comfortable yet dangerous dependence on a fossil-fuel-based economy, toward a more sustainable future for our children and grandchildren.

We owe it to those who come after us.

# | GLOSSARY |

**absorbent boom:** A device used to absorb hydrocarbons, the absorbent boom usually consists of a mesh sleeve filled with polypropylene and looks something like a sausage. Booms can be linked together and are often placed along shorelines as a means of capturing and containing any oil that floats on top of the water.

**accumulator bottle:** A key component of the blowout preventer control unit, its function is to store the nitrogen and hydraulic fluid that operate the BOP. Made of steel and bottle-shaped, the accumulator bottle keeps the gas and liquid under high pressure, typically 3,000 pounds per square inch.

**acoustic system:** Used as an emergency backup for the hydraulically operated system of the blowout preventer in floating drilling rigs that use a subsea BOP stack. The system sends audio signals from transducers on the rig to those on a control pod on the BOP. If communication among the transducers is lost, the system automatically operates BOP components.

**aft:** The area near the stern of a ship.

**air induction:** Intakes on a deck engine that provide air for combustion.

**Anglo-Persian Oil Company:** Founded in the early-20th-century oil fields of modern-day Iran by the British oil speculator William Knox d'Arcy, this company went through several transformations and name changes until 2000, when it became known as BP.

**annular preventer:** A valve device used to seal the annulus around either the drill pipe or other pipe in the blowout preventer. This device is installed in a BOP stack above the block ram blowout preventers.

**annulus:** The space between two solid cylinders. In a well, it is the space between the pipe and the borehole or the pipe and the casing where fluids can flow. Also called annular space.

**Arctic National Wildlife Refuge (ANWR):** A 19.3-million-acre region of northeastern Alaska that is a federally protected national wildlife refuge, 8 millon acres of which are designated as wilderness. The ANWR is home to a wide variety of plant and animal life as well as potentially large reservoirs of oil along its coastal region, known as the 1002. Controversy over whether to drill in the 1002 has persisted since 1977.

**assistant driller:** A member of a drilling rig crew who aids the drillers, sometimes overseeing drilling operations in addition to record keeping.

**ballast:** On semi-submersibles, the water added to the pontoons or bottles in order to submerge them to a specific level. On ships, it is the water placed in tanks to weigh down the craft so that it remains stable and positioned at a proper depth in the water.

**barrel:** The standard measure for petroleum volume in the United States, equal to 42 gallons.

**blind ram:** Used in a blowout preventer as a means of shutting off the wellbore with steel blocks that slide over the hole to seal it off completely. Differs from a pipe ram in that it closes over the entire open space, whereas a pipe ram closes around the pipe to seal off the annulus.

**blind shear rams:** In a blowout preventer, the blind shear rams cut through the drill pipe to form a seal against well pressure. The ends of the rams close against each other, completely isolating the space below.

**block:** A device (traveling block or crown block) used in the rig derrick to lift heavy loads into and out of the hole.

**blowout preventer (BOP):** Strong valves installed at the wellhead to stop uncontrolled flow from the well. On floating offshore rigs, the BOP is located on the seafloor; on jackup or platform rigs, it's on the surface of the water; on land rigs, it's at or slightly below the land's surface.

**blowout preventer panel:** The set of controls used to operate the blowout preventer.

**bottom hole assembly (BHA):** The lower portion of the drill string, made up of the drill bit and other devices, including measured while drilling tools.

**bridge:** The area of a mobile offshore drilling unit from which the captain commands the movement of the rig, including dynamic positioning operations.

**bullheading:** A means of pumping fluid into a well from the surface.

**Bureau of Ocean Energy Management, Regulation and Enforcement (BOEMRE):** Formerly the Minerals Management Service (MMS), this agency within the United States Department of the Interior (DOI) is in charge of regulating oil, gas, and other mineral resources on the outer continental shelf (OCS). Established in 1982 under the Federal Oil and Gas Royalty Management Act, the agency is responsible for issuing lease permits for offshore drilling operations, enforcing safety and environmental regulations, and managing natural resource revenues. It is divided into three agencies: the Bureau of Ocean Energy Management, the Bureau of Safety and Environmental Enforcement, and the Office of Natural Resources Revenue.

**capping stack:** A small blowout preventer used to shut in the Macondo well in July 2010. Built specifically for this purpose, it consisted of one double ram cavity and a single, with choke and kill valves installed that would have allowed flowing the well to the surface had BP chosen to do so.

**captain:** On a floating Transocean drilling rig or mobile offshore drilling unit, the captain is in charge when transporting the rig from location to location. When the rig is drilling, the offshore installation manager is in charge. Also called the master on some rigs. In other drilling companies, the captain is in charge at all times.

**casing:** Steel piping placed inside a wellbore to prevent the wall of the hole from caving in and to flow hydrocarbons to the surface.

**casing hanger:** A circular device used to suspend casing from the wellhead.

**cement plug:** Cement placed inside the wellbore to seal it when abandoning a well, whether temporarily or permanently.

**centralizer:** A device used to center casing in the hole, allowing cement to flow around that casing more evenly.

**chief electronics technician:** The technician in charge of maintaining electronic control systems on an offshore drilling rig.

**chief mate:** The chief mate, who reports to the captain, is generally in charge of safety training and deck operations.

**chief mechanic:** On an offshore drilling rig, the chief mechanic is in charge of maintaining mechanical components such as engines, pumps, and motors.

**choke:** A device consisting of valves and piping used to control the flow of fluids in a line. Chokes have a small opening that restricts the flow of fluids from the hole when a blowout preventer has been closed and a kick is being circulated out of the hole.

**choke line:** A pipe installed below the ram blowout preventers and up the outside of the riser that serves as a conduit for fluids to flow from the well to the surface through the choke manifold.

**choke manifold:** An arrangement of chokes attached to the wellhead through the choke line. When blowout preventers are closed, mud circulates through the choke manifold. The device is also used in well testing as a means of controlling flow from the well.

**containment dome/structure:** A device designed to collect oil flowing from an uncontrolled well on the seafloor.

**control pods:** Devices used to control subsea blowout preventers on floating offshore rigs. Hydraulic fluid and electrical signals flow from the rig into the pods, which then send the signals to the various annular or ram preventers. Two pods, always called the yellow pod and blue pod, are used on subsea BOPs as backup in case one fails.

**control pod receptacles:** The connection between the control pods and the hydraulic system of a subsea blowout preventer.

**Corexit 9527 and 9500:** Controversial oil spill dispersants developed and sold by Nalco Company. Each product contains solvents and chemicals that break up oil into tiny droplets. Designed to apply to surface oil spills, Corexit was applied at the seafloor for the first time in the Macondo spill, with unknown consequences.

**crane operator:** This person operates one of the cranes on a rig that move equipment around the rig floor and between supply vessels and the rig.

**Damon Bankston:** An offshore supply vessel assigned to the *Deepwater Horizon* that played a key role in rescuing the survivors the night of the blowout, April 20, 2010.

**davit:** A device used to lower lifeboats and life rafts from an offshore rig to the water.

**deadman:** An emergency system in the subsea blowout preventer that automatically shuts in the well in the event that hydraulic and electrical control between the rig and the BOP is lost.

**deck:** A floor on a marine vessel.

**deck engine:** A large, high-horsepower engine, usually fueled with diesel, that provides power to an offshore rig.

**deepwater:** At present, *deepwater* refers to offshore drilling operations in 1,000 feet of water or deeper. These depths present a number of special challenges relating to the control of the well, as the blowout preventer sits on the seafloor. This term is relative depending on the capabilities of current technology, thus it refers not to the drilling depth itself, but to the technical complications caused by drilling beyond a certain depth.

**Deepwater Horizon:** The rig that experienced a severe blowout in April 2010 while drilling in the Gulf of Mexico. Owned and operated by Transocean, it was a semi-submersible, dynamically positioned rig capable of drilling ultra-deepwater wells. After drilling the deepest well in history in September 2009, it moved to the Macondo field and began drilling the well that would ultimately blow out and cause the massive Gulf oil spill. Two days after the blowout, the rig sank to the ocean floor.

**deepwater mooring and stabilizing systems:** The means of fixing and steadying a floating offshore rig or other vessel to one place in the water using anchors, cables, and chains.

**derrick:** A large structure on a drilling rig used to support the crown block, traveling block, and drill string. Derricks typically have a pyramid-like shape, with four legs at each corner of the base that converge at the crown block.

**derrick man:** A member of the rig crew who works in the derrick while tripping pipe; during drilling operations, the derrick man generally works in the mud pit area.

**Discoverer Enterprise:** The Transocean-owned drillship sent to perform containment and well-relief operations on site of the *Deepwater Horizon* blowout.

# GLOSSARY

**downhole:** Used to describe something pertaining to the wellbore.

**drawworks:** A large winch that raises and lowers the drill string.

**drill bit:** The very tip of the bottom hole assembly—this is the tool that actually cuts through the rock. Most drill bits work by rotating and grinding.

**drill collar:** Heavy, steel pipe placed between the drill pipe and the bit in the drill stem. The drill collar is used to add weight to the drill bit and create a pendulum effect on the stem, adding to the force of the drill bit.

**driller:** The rig crew foreman. A driller supervises the rig crew and operates the drilling package from the driller's chair in the driller's shack on the rig floor.

**drilling fluid:** Synonymous with *drilling mud*, the term refers to a fluid circulated through the wellbore during rotary drilling operations. It is used to counteract downhole formation pressure, stabilize the borehole, prevent formation swelling, cool the bit, lubricate the drill pipe, control fluid loss, and lift cuttings out of the wellbore and onto the surface. It can be synthetic oil, diesel, or water-based.

**drill pipe:** A high-strength steel pipe that connects the rig surface equipment to the bottom hole assembly. It carries drilling fluid through the bit and allows the rig to control the bottom hole assembly.

**drill pipe tool joint:** The ends of a drill pipe, fitted with threaded components, joining the pipe sections together. The "male" component, on the top part of the pipe, screws into the "female" component at the bottom of the adjoining pipe. Joints are thicker in diameter than the drill pipe itself, friction-welded to the ends of the pipe, and made of a specially heat-treated alloy steel; all these features serve to prevent leaks, sustain the weight of the drill stem, and withstand frequent assembly and disassembly.

**drillship:** A vessel outfitted to hold drilling rig and related equipment. Drillships are held in place by anchors, dynamic positioning, or a combination of the two and are capable of carrying heavy loads and drilling wells in ultra-deepwater. Although they are less stable than semi-submersibles, they can generally bear heavier loads and operate in heavier seas.

**drill string:** The combination of parts, including the drill pipe, drill collars, and bottom hole assembly, used to turn the drill bit at the bottom of the wellbore. It transmits fluid and rotational power from the top drive to the bit.

**drum skimmer:** A steel cylindrical-shaped device used to separate and collect oil that is mixed in water. The motion of the rotating cylinder, or drum, draws in oil, which is then diverted into a collection tank. These machines are generally mobile and float atop the water, but sometimes they exist in fixed units.

**dynamic positioning (DP):** A means of fixing a floating offshore rig in position over an offshore well by using propulsion units instead of mooring anchors. These propulsion units, called thrusters, are located on the hulls of the floating rig and controlled by a computer sensing system that directs and adjusts the motion of the propellers to keep the rig in place. See *fanbeam laser orientation system*.

**dynamic positioning operator (DPO):** The person in charge of monitoring the dynamic positioning system that keeps the rig in its fixed position over the well. The DPO reports to the captain.

**emergency disconnect system (EDS):** In a blowout preventer, the EDS is programmed to close the blind shear rams and separate the riser and lower marine riser package from the well.

**emulsified oil:** Oil mixed with water to the point of gelling, meaning that the water is combined with the oil to the point that it's very difficult to skim without using dispersion methods.

**engine console:** The set of controls used to operate a rig's deck engines.

**engine control room (ECR):** This room contains the monitoring and control devices that operate a rig's deck engines.

**fanbeam laser orientation system:** A system of laser sensors used to operate the dynamic positioning of floating vessels. The system consists of three parts—a laser scanner head, attached by an auto-tilting yoke to the base console, containing special scanning software. The lasers send signals to the thrusters in the hull of the vessel. This method allows for highly accurate remote positioning of vessels.

**fast rescue craft (FRC):** A small rescue craft, usually kept on offshore supply vessels, designed for rapid deployment and rescue of rig personnel who may be in the water.

**fireboat:** A floating vessel outfitted with special equipment designed for fighting fires on ships and along the shoreline. Fireboats have pumps that draw water in from below and nozzles that spray the pumped water out in enormous jets. The mechanisms aboard these boats can pump tens of thousands of gallons of water per minute.

**fire gear locker:** A central location that contains firefighting equipment and clothing.

**fire team:** The rig crew trained for firefighting duty on a ship or offshore rig.

**fishing:** The process of removing pieces of equipment or metal debris left inside a wellbore during prior drilling or extraction operations. Such debris must be dislodged before new operations can begin.

**flag nation:** The country of registration of ships and mobile offshore drilling units.

**flex joint flange:** A connector at the top of a subsea blowout preventer that allows movement of the riser connected to a floating drilling rig. Usually containing an elastomer, the flex joint allows the riser to move more easily with the ocean currents.

**flex trend:** The region of the offshore between shallow and deepwater, generally in depths between 600 and 1,500 feet. It is where the shallow continental shelf ends and begins to slope downward into deepwater.

**floater:** Slang for a floating drilling rig or drillship.

**floating rig:** A mobile offshore drilling unit that floats atop the water and is not fixed to the seafloor. Anchors may hold the floating rig to the seafloor, but when dynamic positioning is used, the unit may not contact the seafloor at all. Types of floating rigs include semi-submersibles, barge rigs, and drillships.

**floor crew:** The workers on a drilling rig who are assigned to the rig floor and report to the driller and assistant driller.

**floor hand:** This member of the rig crew works on the rig floor and reports to the driller and assistant driller.

**flow line:** 1. In drilling, the flow line is the pipe that directs the drilling fluid (or mud) coming from the top of the wellbore to the mud storage and treatment tanks at the surface. 2. In production, the flow line is the pipe that directs the oil from the well to storage or processing equipment.

**flowmeter:** A gauge that records the movement of fluids. It's used in blowout preventer panels to show movement of hydraulic fluid during annular and ram operation.

**flow rate:** The volume of oil flowing per unit of time. Flow rate is generally expressed as barrels per minute or barrels per day.

**Flow Rate Technical Group:** The group of scientists and engineers assembled by the National Incident Command to provide an independent scientific estimate of oil flow rates in the Gulf. Headed by Marcia McNutt of the United States Geological Survey, the group comprised four subteams that each took a different approach to calculating flow rates, with the aim of converging upon the most accurate estimate possible. The subteams were the Plume Modeling Team, the Nodal Analysis Team, the Mass Balance Team, and the Resevoir Modeling Team.

**formation:** A subsurface layer of rock.

**frac/frac job:** A means of enhancing well production by using hydraulic pumps to force a specially blended fluid at high pressure to create cracks in the subsurface formation around the well. The fluid contains particles such as sand and aluminum pellets that serve to hold open the cracks, so that once the hydraulic pressure is removed, the cracks remain open and oil can flow through.

**frac gradient:** The ratio determining the pressure needed to fracture a formation as a function of well depth, measured as psi/foot.

**gas sensor:** Attached at the flow line during drilling, it is used to detect formation gas in the drilling mud coming from the wellbore.

**GDMSS (Global Maritime Distress System) button:** An emergency device that transmits the precise location and condition of a vessel in distress.

**general alarm:** A vessel-wide alarm that alerts the crew to a dangerous condition.

**heave compensator:** Used to correct for the effects of ship movement on a floating rig, this device prevents the drilling equipment from being lifted and dropped due to sea motion.

**helideck:** The landing platform on an offshore drilling rig. On the *Deepwater Horizon*, it was located on the roof of the bridge.

**hot stab:** A device inserted into a port on a subsea blowout preventer by a remotely operated vehicle to initiate operations by injecting hydraulic fluid.

**hydrates:** Compounds of water and hydrocarbons resembling ice crystals that occur in situations where natural gas encounters water at low temperature and high pressure.

**hydraulic system:** A system controlled with pressurized fluids rather than electronics. Subsea blowout preventer control systems are typically hydraulically controlled.

**hydrocarbon:** Organic compounds made up of hydrogen and carbon. They can be simple gases like methane, liquids such as oil and gas, or solids like coal. Petroleum is a complex hydrocarbon mixture.

**hydrostatic pressure:** The pressure exerted by a column of fluid at rest, measured in pounds per square inch (psi). Hydrostatic pressure increases with water depth and is relevant in drilling as a means of measuring the pressure of the drilling mud in the wellbore.

**inflatable boom:** Used instead of traditional foam-filled absorbent booms, these cylindrical-shaped tubes form a snakelike physical barrier to contain oil spills on the water's surface. They are easier to transport than absorbent booms and are often used in open waters, where their higher buoyancy-to-weight ratio makes them more adaptable to wave movement.

**jackup rig:** An offshore drilling rig designed with legs that sit on the seafloor, supporting the rig above the surface of the water.

**Joint Investigation:** An investigation into the Macondo well blowout and surrounding events being conducted by the US Coast Guard and the Bureau of Ocean Energy Management, Regulation and Enforcement (BOEMRE, formerly the Minerals Management Service).

**junk shot:** A strategy designed to stop the flow of a well by pumping blocking agents such as rubber, knotted robes, golf balls, and other solid materials to bridge over leaks.

**kick:** An influx of fluids into the wellbore that occurs when fluid from the pores of a rock formation (which can include water, gas, or oil) flows in as a result of pressure in the wellbore being lower than that of the surrounding formation. These fluids can then travel up through the wellbore, with the upward pressure risking a blowout if not quickly addressed. A kick can happen when the hydrostatic pressure of the drilling fluid (mud) is too low to prevent the flow of formation fluid into the well. Signs of a kick occurring are a sudden increase in drilling rate, a gain in the pits, and an increase in gas in the drilling mud.

**kick prevention:** The steps taken during the drilling of a well to ensure that the wellbore exerts sufficient pressure to prevent a kick, which could in turn cause a blowout. Kick prevention includes accurate gauging of pore pressure and fracture pressure gradients, use of sound drilling and casing equipment, proper drilling mud quality and weight, and correct operational practices.

**kill:** The steps taken to control a kick. These include using blowout preventers to shut the well in at the surface, circulating the kick fluid out of the wellbore, and replacing the mud with sufficiently heavy kill mud.

**kill calculations:** Determining the mud weight and density needed in order to execute a successful kill and halt the flow of kick fluid.

**kill fluid:** Mud that is heavy enough to stop the flow of kick fluid from the rock formation in a well hole.

**kill line:** When drilling mud cannot be pumped into a well through the drill string due to activation of the blowout preventer (or because there's no drill pipe in the hole), the kill line serves as a conduit through which heavier kill mud can be bullheaded into the well to stop the flow of kick fluid. The kill line is connected to the BOP stack below the ram preventers.

**KIP:** A measure of force equal to 1,000 pounds.

**knots:** A measure of speed; 1 knot equals approximately 1.15 miles per hour.

**latching collet connector:** A hydraulically controlled connector that attaches the blowout preventer stack to the wellhead.

**line:** A pipe that serves as a conduit for liquid flows.

**liner:** A length of casing that extends from the bottom of the hole to just inside the next largest string of casing. The top of the liner is connected to the casing above by a liner hanger that seals the casing from outside formation pressure by forming a barrier to any influx of wellbore fluids.

**liner hanger:** See *liner.*

**lockdown sleeve:** A device used to hold down the casing hanger and production casing. If pressure conditions are such that the casing exerts upward pressure, the lockdown sleeve prevents it from rising up in the wellhead, a condition that could allow hydrocarbons to escape into the riser and make their way to the surface.

**logging while drilling (LWD):** Measuring properties of the formation—including resistivity, porosity, and sonic velocity—while a hole is being drilled. LWD equipment is integrated into the lower portion of the drill string.

**long string design:** The longest string of casing that runs near or through the producing zone of the well to the surface. It is the last string of casing to be set in place in the drilling process and is differentiated from a liner in that its only barriers are the cement at the bottom of the casing and the casing hanger at the surface.

**loop current:** A current of warm water that flows clockwise through the Gulf of Mexico, carrying water through the Gulf, through the straits between Florida and Cuba, and out into the Atlantic Ocean.

**lost circulation zone:** A subsurface rock formation that allows drilling mud to escape from the wellbore rather than circulating back to the surface because of weakness or fractures.

**lower marine riser package (LMRP):** The upper segment of the subsea blow-out preventer stack, the LMRP includes the blowout LMRP connector, annular preventers, the flex joint, and the marine riser connector. (Also see *latching collet connector.*)

**manifold:** 1. A system of piping or valves designed to control or reroute a flow. 2. A pipe with side outlets allowing it to connect to other pipes.

**man overboard drills:** Regular emergency practice held to train crew members to rescue those who have fallen off a vessel into the water.

**maximum riser angle:** The angle at which a deepwater rig will disconnect from a subsea blowout preventer to keep from damaging the riser or subsea equipment.

**maximum working pressure:** The maximum allowed pressure that can be applied to a wellhead component or pipe before damage occurs.

**mobile offshore drilling unit (MODU):** An offshore drilling rig that is mobile via flotation, although it's not necessarily floating during operation. MODUs include jackups, drillships, semi-submersibles, and submersibles.

**moonpool:** The opening in the hull of an offshore drilling vessel that allows drilling equipment to pass through.

**mooring system:** The anchors, cables, and chains used to fix a floating offshore vessel in place.

**motor vessel (MV):** A ship propelled by an internal combustion engine.

**mud:** See *drilling fluid.*

**mud engineer:** The individual responsible for ensuring that the drilling fluid used on a rig maintains the correct properties for the job. Duties include frequent testing of the mud as well as prescribing treatments to make needed adjustments. The mud engineer is typically employed by the drilling fluid supply company and works closely with the rig supervisor and derrick man.

**mudline:** The seafloor.

**mud properties:** The characteristics of drilling fluid that affect its performance. These include density, or the mass per unit volume; viscosity, or resistance to flow; fluid loss control, or resistance to seepage into the rock formation; sand content; and gel strength, or the ability to hold suspended particles.

**natural gas liquids (NGL):** Natural gas components that are liquid at the surface in field facilities or gas-processing plants. They include propane, butane, pentane, hexane, and heptane, and are classified according to their vapor pressure.

**Notice to Lessees (NTL):** Notice from the BOEMRE (MMS) to communicate directives from the agency to oil and gas operators on federal lands.

**offshore installation manager (OIM):** The seniormost position on a mobile offshore drilling unit. Specially trained and certified, the OIM is in charge of all operations on the MODU and is responsible for the safety of all personnel. Toolpushers report to the OIM.

**offshore supply vessel (OSV):** A ship designed to transport goods to and from the offshore oil rig.

**oil boom:** A floating device used to contain an oil spill. Oil booms are usually linked together in a sausagelike formation to form a physical barrier that prevents oil from spreading on the surface. They can be either absorbent or nonabsorbent. (Also see *absorbent boom, floating boom.*)

**oil plume:** A subsurface concentration of hydrocarbons—generally dispersed oil—from deepwater well flows. Locating and identifying plumes has proven difficult, and the damage they cause is as yet unknown.

**oil spill remediation:** The process of cleaning up an oil spill and attempting to return the environment to its pre-spill condition.

**outer continental shelf (OCS):** Offshore regions subject to federal jurisdiction, usually referring to mineral rights ownership regulations. The OCS begins where the state's mineral rights ownership boundaries end and extends out until depths exceed the limits for feasible exploration, generally at about 8,000 feet.

**painter line:** A line attached to the bow of a boat and used for docking, towing, or tying the vessel to another object.

**particle image velocimetry:** A means of obtaining data about the properties of a fluid using floating particles.

**pipe ram:** A device used in high-pressure split-seal blowout preventers to fit around the drillpipe and form a seal. There is a semicircular hole on the edge of a pipe ram meant to fit half the circumference of a pipe; this is matched with an identical ram horizontally to form a complete seal. However, because pipe rams typically fit only a certain size of pipe and cannot close around tool joints or drill collars, they are increasingly being replaced by variable-bore rams, which can adjust to fit a range of pipe circumferences.

**platform:** A structure that houses all the necessary components for the drilling, extraction, processing, and transportation of oil and/or natural gas from offshore well locations, including workers and machinery. Platforms are stationary but not necessarily fixed—many are mobile and kept in place by dynamic positioning and/or mooring systems. Fixed platforms are anchored to the seafloor with concrete or steel legs and are used for drilling in shallower depths. Deepwater platforms are typically mobile.

**pod:** See *control pod*.

**pontoon:** A flat-bottomed structure submerged beneath the ocean's surface and used to support a semi-submersible drilling rig. Structural columns connect pontoons with the operating decks, which are located above the surface.

**pore pressure:** The force of the fluids in the openings in a rock or mass of rocks, pore pressure is usually measured in hydrostatic pressure.

**pore space:** A void or space within a rock or group of rocks that can fill with air, water, or hydrocarbons.

**port:** The left side of the ship from the perspective of an onboard person facing the bow.

**rack-and-pinion system:** A gear consisting of a bar with teeth (the rack) that fits to a smaller rotating gear (the pinion).

**ram block:** The steel block device that closes off the wellbore in ram-type blowout preventers. These are hydraulically operated and always meet in horizontally opposing pairs to form the seal over the well opening.

**ram cavity:** The area inside a ram blowout preventer where the rams and their operating parts move upon activation of the BOP.

**ram preventer:** A type of blowout preventer that uses ram blocks to seal off the wellbore and cut the flow of pressure coming from the well.

**remotely operated vehicle (ROV):** In deepwater operations, an unmanned vehicle sent down to inspect and manipulate subsea machinery such as the blowout preventer. ROVs are controlled by personnel on the surface and are connected via a series of cables called a tether, which transmits video, electrical, and other data signals between the subsea robot and the platform.

**rig:** Technically, the machinery used to drill a well, including mud tanks and pumps, the derrick or mast, the drawworks, the drill string, and the topdrive. Offshore drilling platform outfits that include housing for personnel are often referred to as rigs. Also called the drilling package.

**riser:** In offshore drilling operations, the riser connects the well at the seafloor with the surface drilling equipment. It's a series of pipe sections connected with special components to adjust for sea motion. The pipe guides the drill string down into the well and serves as a conduit for drilling fluid coming up from the wellbore to the surface for processing.

**riser angle:** See *maximum riser angle.*

**riser insertion tool (RIT):** This repair tool designed by BP comprises a siphoning steel pipe extending from a drillship to the leaking riser gushing on the ocean floor.

**roustabout:** A member of the rig crew and generally considered an entry-level position, a roustabout works at the direction of the driller or crane operator.

**rubber cementing plugs:** The rubber plugs used in cementing operations prevent contamination between drilling fluid and the cement. The bottom plug is run before the cement to clear the casing walls of any other fluids; the top plug is run immediately after the cement to both clean the cement off the walls and prevent drilling fluid from mixing in with the cement.

**sand berm:** A barrier made of sand, used during oil spills to protect the coast from oil washing ashore.

**seal assembly:** This device seals the annulus between each joint in a casing string, closing off each casing section from the higher pressure of the section below it.

**semi-submersible rig:** A floating mobile offshore drilling unit in which the base structures—either steel columns attached to pontoons or giant bottle-shaped steel cylinders—are submerged below the ocean's surface while the operating deck hovers above the water. By minimizing contact with the wave and wind variability on the surface, these vessels achieve greater stability than other floating units such as drillships and barges. Either self-propelled or towed to an offshore location and kept in place by anchored mooring systems, dynamic positioning mechanisms, or a combination, these rigs are ideal for conducting operations in rough seas and deepwater.

**separator:** A spherical or cylindrical drum that uses gravity to separate oil and gas from water (or mud) in a well output stream, wherein the heavier fluid settles to the bottom and the lighter fluid rises to the top.

**shear out:** An action that activates a downhole tool by pulling or setting down weight. This shears a metal pin that otherwise holds the tool in neutral.

**shear ram:** See *blind shear rams.*

**shut in:** Stopping a well from flowing with a valve or BOP.

**skimming:** A process of removing oil from water wherein water containing small quantities of oil flows into a tank; when gravity separates the substances and the oil rises to the top, skimmer blades inside the tank send the oil layer to a discharge line. Skimmer tanks are effective only when the water has relatively small quantities of oil in it.

**spacer fluid:** A liquid used between two other special-purpose liquids as a means of displacing one liquid so it can be replaced by the other, often used when changing from one type of drilling mud to another, or during cementing operations to separate drilling mud from cement. Spacer fluids must be compatible with and benign to both special-purpose liquids, and they can be oil- or water-based, depending on the situation. A spacer fluid used to displace drilling mud might be thickened with fly ash or barite so that its weight and viscosity will be sufficient to push out the thick drilling mud.

# GLOSSARY

**spinning chain:** In the assembly of a drill pipe, this is a length of chain wrapped around one pipe joint as it is being connected to a new section of pipe, then pulled such that it causes the new drillpipe joint to function as a spool, spinning into the screws of the original pipe. This manual method is highly dangerous and rarely used nowadays; other, safer spinning devices have replaced it.

**splash:** Used as a verb, this refers to running and setting the blowout preventer on the subsea wellhead.

**spool:** A length of pipe with flanges welded on each end that allow the pipe to be bolted tó a wellhead or production vessel.

**stack:** The vertical arrangement of blowout preventer components.

**standby generator:** A piece of emergency equipment designed to restore power to a rig in case the regular deck engines and generators fail.

**starboard:** The right side of a ship from the perspective of a person onboard facing the bow.

**static kill procedure:** Pumping heavy drilling mud down the well through choke and kill lines into the blowout preventer in an effort to kill the well completely. This is similar to the top kill procedure but is performed on a capped well that is no longer gushing oil and thus requires lower rates and pressures for pumping.

**subsea BOP:** Used in offshore drilling rigs, this is a blowout preventer located on the seafloor.

**substrata:** Layers of the earth's crust that lie below the surface layer of soil or seafloor. These contain high levels of fossilized organic matter and are a rich source of hydrocarbon mineral resources. Drilling rigs penetrate the substrata in order to extract petroleum and other hydrocarbons.

**supershear:** A ram preventer designed to cut casing.

**supertanker:** A tanker ship that can transport over 100,000 deadweight tons (DWT), which is the measure of all added weight on the ship including cargo, provisions, fuel, ballast, fresh water, passengers, and crew. Supertankers with a capacity of 500,000 DWT or less are called very large crude carriers; if their capacity is greater than 500,000 DWT, they're known as ultra-large crude carriers. (Also see *tanker*.)

**surface vessel:** A ship that floats along the surface of the sea and cannot submerge.

**synthetic drilling mud:** A drilling fluid that uses a synthetic oil as its base rather than water or diesel oil.

**tanker:** Also called a tank ship, this floating vessel is designed to transport oil, liquefied petroleum gas (LPG), liquefied natural gas (LNG), or substitute natural gas (SNG). Tankers with capacities of greater than 100,000 deadweight tons are called supertankers.

**test ram:** A ram preventer designed to hold pressure from the top of the ram rather than below. It's the bottom ram in a subsea blowout preventer stack used to test the other preventers in the stack.

**thruster system:** In a dynamic positioning system, this propulsion device located on the hull of a floating rig vessel uses a computerized control and sensing system to keep the vessel stationary.

**tongs:** Large pipe wrenches used to assemble or disassemble downhole equipment such as drill pipe, tubing, or casing. Hydraulically or pneumatically operated turning wrenches are called power tongs.

**toolpusher:** Manager of the rig crew, the toolpusher reports to the offshore installation manager.

**topdrive:** A hydraulically powered swivel that hangs from the traveling block in the derrick and rotates the drill string.

**top hat:** Designed by BP, this small containment cap was lowered over the Macondo well in June 2010. The effort was partially successful, collecting as much as half of the well flow.

**top kill procedure:** A plan to overcome the flow of oil gushing from the Macondo well by pumping heavy drilling mud into the wellhead at high rates and high pressure, then sealing the well with cement. The procedure failed on May 29, 2010.

**topside:** The upper deck on a vessel.

**total depth (TD):** The end point of a well as measured by the length of pipe used to reach the bottom.

**tours:** An oil field term for a shift. Most offshore rigs run in 12-hour shifts.

**tripping pipe:** Running pipe into, or pulling pipe out of, a well.

**trophic cascade:** A collapse of an ecosystem, usually caused by outside forces; it's generally thought of as a bottom-up degradation of the food chain but is observed to be much more complex when it occurs.

**umbilical line:** A control line that runs from the rig on the surface to the blowout preventer on the seafloor.

**Unified Command:** A government organization established after the blowout of the Macondo well and comprising multiple federal agencies as well as the private companies involved in the incident.

**USGS:** United States Geological Survey.

**variable-bore ram:** A ram preventer that can close around different sizes of pipe.

**variable choke:** An adjustable choke that can vary the flow of a well.

**Vessels of Opportunity:** A program begun by BP that employed private fishing boats to assist in cleaning up the oil spill. At its peak, it employed up to 3,000 vessels.

**Wärtsilä deck engine:** A large diesel engine (manufactured by Wärtsilä Corporation of Finland) that powers ships and offshore rigs.

**weir skimmer:** A skimmer that collects oil from water by running the mixture over an adjustable fence; the oil can flow over, while the water is held back.

**wellbore:** The hole being drilled.

**wellhead:** The assembly on the top of a well that contains the production casing and tubing.

**well site leader:** The operator's representative on an offshore rig. Generally called the company man, the well site leader is in charge of drilling and completing the well.

**wet trees:** Subsea wellheads.

**wire rope:** Steel cable used in hoisting operations on the rig.

**wire saw:** A saw used for precision cuts on subsea piping, controlled by an ROV.

# | APPENDIX |

# Diagrams and Images from the Blowout

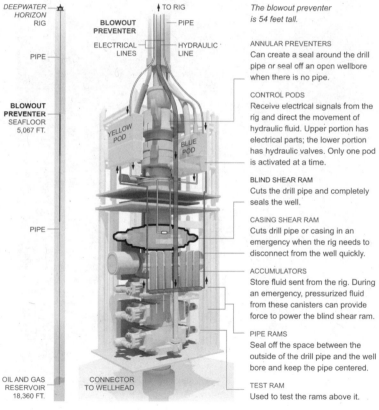

DEEPWATER HORIZON RIG

TO RIG

BLOWOUT PREVENTER

PIPE

ELECTRICAL LINES

HYDRAULIC LINE

PIPE

BLOWOUT PREVENTER SEAFLOOR 5,067 FT.

YELLOW POD

BLUE POD

PIPE

OIL AND GAS RESERVOIR 18,360 FT.

CONNECTOR TO WELLHEAD

*The blowout preventer is 54 feet tall.*

ANNULAR PREVENTERS
Can create a seal around the drill pipe or seal off an open wellbore when there is no pipe.

CONTROL PODS
Receive electrical signals from the rig and direct the movement of hydraulic fluid. Upper portion has electrical parts; the lower portion has hydraulic valves. Only one pod is activated at a time.

BLIND SHEAR RAM
Cuts the drill pipe and completely seals the well.

CASING SHEAR RAM
Cuts drill pipe or casing in an emergency when the rig needs to disconnect from the well quickly.

ACCUMULATORS
Store fluid sent from the rig. During an emergency, pressurized fluid from these canisters can provide force to power the blind shear ram.

PIPE RAMS
Seal off the space between the outside of the drill pipe and the well bore and keep the pipe centered.

TEST RAM
Used to test the rams above it.

*New York Times* Graphics

A cutaway view of the typical subsea blowout preventer. Visible are the ram preventers on the bottom of the stack, the annular preventers at the top, and the control system. The blowout preventer is the last line of defense to prevent loss of control of a well.

The *Deepwater Horizon*'s blowout preventer, on the deck of the Q4000, after being recovered from the seafloor above BP's blown-out Macondo well. The failure of the blowout preventer was central to the disaster and became the focus of the ensuing investigation. The US Department of Justice took possession of the blowout preventer immediately after it was recovered.

## Relief Well Diagram

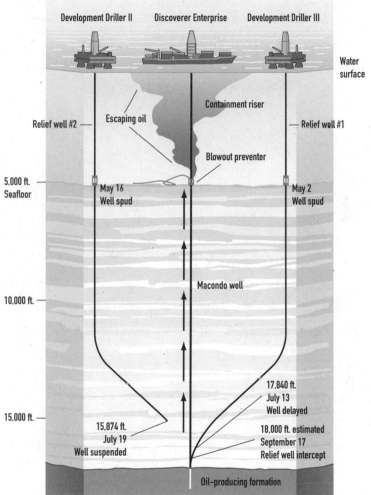

Illustration by Phil Laughlin Studio

Diagram of the two relief wells drilled to kill the blown-out Macondo well. Relief well #1 successfully intercepted the blown-out well on September 17, 2010, and the kill was declared successful two days later. Relief-well drilling was delayed for almost two months due to decisions made by the US government and BP, and weather delays. Had the relief-well strategy not been second-guessed, the well could have been killed much earlier.

## Macondo Well Oil Spill

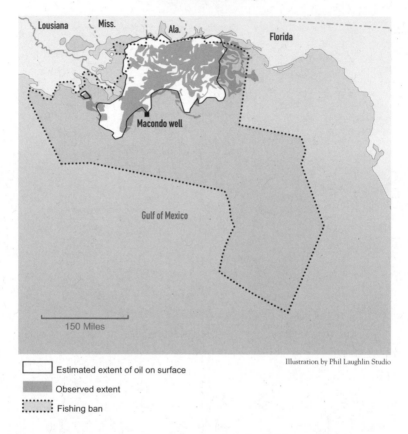

Illustration by Phil Laughlin Studio

☐ Estimated extent of oil on surface

▨ Observed extent

⋯ Fishing ban

Surface oil expanded over the Gulf in the months following the BP disaster. Of the 4.9 million barrels of oil that government officials estimate flowed into the Gulf of Mexico, only 800,000 barrels were captured. Some independent scientists project higher spill volumes. But all agree that the spill was by far the largest in US history. It was also the first to leave both surface oil and subsea plumes in its wake. The map above does not reflect the plumes. It depicts the surface oil estimates from the National Oceanic and Atmospheric Administration as well as the observed coverage, based on satellite and aerial images. Adapted from the *New York Times* interactive map, "Tracking the Oil Spill in the Gulf," http://www.nytimes.com/interactive/2010/05/01/us/20100501-oil -spill-tracker.html

*Left:* The self-propelled *Deepwater Horizon* en route to its drilling location. The *Deepwater Horizon,* a mobile offshore drilling unit, or MODU, operated as a floating rig before being destroyed during BP's Macondo well blowout on April 20, 2010.

*Below:* The *Deepwater Horizon* on location. To provide a stable drilling platform, a semi-submersible rig's pontoons are submerged below the surface of the water to minimize the effects of winds, waves, and currents. The *Horizon* had drilled in the Gulf of Mexico in water deeper than 9,000 feet.

Photos courtesy of Transocean

# | ENDNOTES |

## Chapter 1. Unleashing the Beast and Paying the Price

1. Deepwater Horizon Joint Investigation Committee, "Investigation of Deepwater Horizon Explosion, Douglas Brown Testimony, May 26, 2010," C-Span Video Library, www.c-spanvideo.org/program/id/225009 (accessed September 10, 2010).

2. CNN, "Oil Flow Estimate Increased; Should President Obama Get Tough with BP?" *The Situation Room with Wolf Blitzer,* June 11, 2010, archives.cnn.com/ TRANSCRIPTS/1006/11/sitroom.02.html (accessed September 10, 2010).

3. Deepwater Horizon Joint Investigation Committee, "Investigation of Deepwater Horizon Explosion, Mike Williams Testimony, June 24, 2010," C-Span Video Library, www.c-spanvideo.org/program/294728-1 (accessed September 10, 2010).

4. CNN, "The Situation Room: BP Releases Video of Leak Source," *The Situation Room with Wolf Blitzer,* May 11, 2010, archives.cnn.com/ TRANSCRIPTS/1005/11/ltm.02.html (accessed September 10, 2010).

5. Deepwater Horizon Joint Investigation Committee, "Deepwater Horizon Joint Hearing Testimony Videos for July 19, 2010," Steve Bertone testimony, www .deepwaterinvestigation.com/go/doc/3043/796615 (accessed September 10, 2010).

6. Deepwater Horizon Joint Investigation Committee, "Investigation of Deepwater Horizon Explosion, Christopher Pleasant Testimony, May 28, 2010," C-Span Video Library, www.c-spanvideo.org/program/293776-2 (accessed September 10, 2010).

7. Deepwater Horizon Joint Investigation Commitee, "Investigation of Deepwater Horizon Explosion, David Young Testimony, May 27, 2010," C-Span Video Library, www.c-spanvideo.org/program/293757-3 (accessed September 10, 2010).

8. Deepwater Horizon Joint Investigation Committee, "Investigation of Deepwater Horizon Explosion, Curt Kuchta Testimony, May 27, 2010," C-Span Video Library, www.c-spanvideo.org/program/id/225567 (accessed September 10, 2010).

9. Douglas A. Blackmon et al., "There Was 'Nobody in Charge,'" *Wall Street Journal,* May 27, 2010, online.wsj.com/article/SB10001424052748704113504575264721101985024.html (accessed September 10, 2010).

10. Deepwater Horizon Joint Investigation Committee, "USCG/MMS Marine Board of Investigation into the Marine Casualty, Explosion, Fire, Pollution, and Sinking of Mobile Offshore Drilling Unit Deepwater Horizon, with Loss of Life in the Gulf of Mexico April 21–22, 2010," Joint Investigation official website, testimony of Anthony Gervasio, May 11, 2010, www.deepwaterinvestigation .com/external/content/document/3043/621903/1/Deepwater%20Horizon%20 Joint%20Investigation%20Transcript%20-%20May%2011 (accessed September 10, 2010).

11. Deepwater Horizon Joint Investigation Committee, "USCG/MMS Marine Board of Investigation into the Marine Casualty, Explosion, Fire, Pollution, and Sinking of Mobile Offshore Drilling Unit Deepwater Horizon, with Loss of Life in the Gulf of Mexico April 21–22, 2010," Joint Investigation official website, testimony of Alwin Landry, May 11, 2010, www.deepwaterinvestigation

.com/external/content/document/3043/621903/1/Deepwater%20Horizon%20
Joint%20Investigation%20Transcript%20-%20May%2011 (accessed September
10, 2010).

12. US Energy Information Administration Independent Statistics and Analysis,
"Gulf of Mexico Fact Sheet," September 3, 2010, www.eia.doe.gov/special/
gulf_of_mexico/index.cfm (accessed September 13, 2010).

13. Brett Clanton, "Deep-Water Ban Sending Ripples Through Houston," *Houston Chronicle*, June 6, 2010.

## Chapter 2. The Well from Hell: Drilling the Deepwater

1. US House of Representatives Committee on Energy and Commerce, "Macondo Prospect Well Information," Committee on Energy and Commerce website, September 2009, energycommerce.house.gov/documents/20100614/Macondo .Prospect.Well.Information.pdf (accessed September 10, 2010).

2. Bureau of Ocean Energy Management, Regulation and Enforcement, "Central Gulf of Mexico Planning Area Lease Sale 206 Information," BOEMRE website, March 19, 2008, www.gomr.boemre.gov/homepg/lsesale/206/cgom206.html (accessed September 10, 2010).

3. SubseaIQ, "Offshore Field Development Projects: Macondo," SubseaIQ website, www.subseaiq.com/data/Project.aspx?project_id=562 (accessed September 10, 2010).

4. Rigzone, "Rig Data: Deepwater Horizon," Rigzone website, rigzone.com/data/ rig_detail.asp?rig_id=153 (accessed September 10, 2010).

5. Ibid.

6. Rigzone, "BP Makes Giant Deepwater Discovery with Tiber," Rigzone website, rigzone.com/news/article.asp?a_id=79913 (accessed September 17, 2010).

7. Transocean, "Fleet Specifications: Transocean Marianas," Transocean website, deepwater.com/fw/main/Transocean-Marianas-77C16.html?LayoutID=17 (accessed September 18, 2010).

8. SubseaIQ, "Offshore Field Development Projects: Macondo," SubseaIQ website, www.subseaiq.com/data/Project.aspx?project_id=562 (accessed September 10, 2010).

9. Deepwater Horizon Joint Investigation Committee, "Mark Hafle Testimony, May 28, 2010," C-Span Video Library, www.c-spanvideo.org/program/293776-1 (accessed September 10, 2010).

10. Mike Williams and Scott Pelley, "Blowout: The Deepwater Horizon Disaster," *60 Minutes*, CBS News (May 16, 2010), www.cbsnews.com/ stories/2010/05/16/60minutes/main6490197.shtml (accessed September 10, 2010).

11. US Senate Committee on Finance, "Interior Department Strengthens Blowout Prevention Requirements After Grassley Inquiry," US Senate Committee on Finance website, June 18, 2010, finance.senate.gov/newsroom/ranking/ release/?id=41e766b2-c848-4750-b01c-a2754ee54ced (accessed September 10, 2010).

12. Deepwater Horizon Joint Investigation Committee, "Investigation of Deepwater Horizon Explosion, Douglas Brown Testimony, May 26, 2010," C-Span Video Library, www.c-spanvideo.org/program/id/225009 (accessed September 10, 2010).

13. Deepwater Horizon Joint Investigation Committee, "Investigation of Deepwater Horizon Explosion, Mike Williams Testimony, June 24, 2010," C-Span Video Library, www.c-spanvideo.org/program/294728-1 (accessed September 10, 2010).

14. Deepwater Horizon Joint Investigation Committee, "Deepwater Horizon Joint Hearing Testimony Videos for July 19, 2010," Steve Bertone testimony, www .deepwaterinvestigation.com/go/doc/3043/796615 (accessed September 10, 2010).

15. Deepwater Horizon Joint Investigation Committee, "Investigation of Deepwater Horizon Explosion, Christopher Pleasant Testimony, May 28, 2010," C-Span Video Library, www.c-spanvideo.org/program/293776-2 (accessed September 10, 2010).

16. Deepwater Horizon Joint Investigation Commitee, "Investigation of Deepwater Horizon Explosion, David Young Testimony, May 27, 2010," C-Span Video Library, www.c-spanvideo.org/program/293757-3 (accessed September 10, 2010).

17. Deepwater Horizon Joint Investigation Committee, "Investigation of Deepwater Horizon Explosion, Curt Kuchta Testimony, May 27, 2010," C-Span Video Library, www.c-spanvideo.org/program/id/225567 (accessed September 10, 2010).

18. Douglas A. Blackmon et al., "There Was 'Nobody in Charge,'" *Wall Street Journal*, May 27, 2010, online.wsj.com/article/SB10001424052748704113504575264721101985024.html (accessed September 10, 2010).

19. Deepwater Horizon Joint Investigation Committee, "USCG/MMS Marine Board of Investigation into the Marine Casualty, Explosion, Fire, Pollution, and Sinking of Mobile Offshore Drilling Unit Deepwater Horizon, with Loss of Life in the Gulf of Mexico April 21–22, 2010," Joint Investigation official website, testimony of Anthony Gervasio, May 11, 2010, www.deepwaterinvestigation .com/external/content/document/3043/621903/1/Deepwater%20Horizon%20Joint%20Investigation%20Transcript%20-%20May%2011 (accessed September 10, 2010).

20. Deepwater Horizon Joint Investigation Committee, "USCG/MMS Marine Board of Investigation into the Marine Casualty, Explosion, Fire, Pollution, and Sinking of Mobile Offshore Drilling Unit Deepwater Horizon, with Loss of Life in the Gulf of Mexico April 21–22, 2010," Joint Investigation official website, testimony of Alwin Landry, May 11, 2010, www.deepwaterinvestigation .com/external/content/document/3043/621903/1/Deepwater%20Horizon%20Joint%20Investigation%20Transcript%20-%20May%2011 (accessed September 10, 2010).

## Chapter 3. How Blowout Preventers Don't Work

1. American Society of Mechanical Engineers, "Cameron First Ram-Type BOP: An ASME Engineering Landmark," Cooper Cameron Corporation Division, July 14, 2003, files.asme.org/asmeorg/Communities/History/Landmarks/9570.pdf.

2. Det Norske Veritas, "Energy Report: Beaufort Sea Drilling Risk Study (Report to Transocean, July 31, 2009)," Louisiana State University Library, www.lib.lsu.edu/ref/oilspill/BOP_failure_study_documents.pdf, page 23.

3. West Engineering Services, "Shear Ram Capabilities Study for US Minerals Management Service (Report to the Mineral Mangement Service, July 2004)," Lousiana State University Library, www.lib.lsu.edu/ref/oilspill/BOP_failure_study_documents.pdf.

4. Christopher S. Young, "Letter Agreement for Conversion of VBR to a Test Ram," US House of Representatives Committee on Energy and Commerce website, October 11, 2004, http://energycommerce.house.gov/documents/20100617/Transocean.Letter.BOP.Modifications.pdf (accessed September 16, 2010).

5. Deepwater Horizon Joint Investigation Committee, "Investigation of Deepwater Horizon Explosion, Christopher Pleasant Testimony, May 28, 2010," C-Span Video Library, www.c-spanvideo.org/program/293776-2 (accessed September 10, 2010).

6. CNN, interview of Tyrone Benton by Drew Griffin on *Anderson Cooper 360°*, June 22, 2010, CNN Transcripts, http://archives.cnn.com/TRANSCRIPTS/1006/22/acd.01.html (accessed September 16, 2010).

7. Hilary Andersson, "BP Was Told of Oil Safety Fault 'Weeks Before Blast,'" *BBC News*, June 21, 2010, www.bbc.co.uk/news/10362139 (accessed September 10, 2010).

8. Deepwater Horizon Joint Investigation, "Investigation of Deepwater Horizon Explosion, Ronald Sepulvedo Testimony, July 20, 2010," C-Span Video Library, www.c-spanvideo.org/program/294623-5 (accessed September 10, 2010).

9. Deepwater Horizon Joint Investigation, "Investigation of Deepwater Horizon Explosion, Harry Thierens Testimony Part 1, August 25, 2010," C-Span Video Library, www.c-spanvideo.org/program/295172-2 (accessed September 10, 2010).

10. US Department of the Interior Press Office, "Interior Issues Directive to Guide Safe, Six-Month Moratorium on Deepwater Drilling," US Department of the Interior website, www.doi.gov/news/pressreleases/Interior-Issues-Directive-to-Guide-Safe-Six-Month-Moratorium-on-Deepwater-Drilling.cfm (accessed September 10, 2010).

11. Associated Press, "Judge Blocks Gulf Offshore Drilling Moratorium; White House Will Appeal," *New Orleans Times-Picayune* website, June 22, 2010, http://www.nola.com/news/gulf-oil-spill/index.ssf/2010/06/judge_blocks_gulf_offshore_dri.html (accessed September 16, 2010).

12. US Department of the Interior, "MMS Deepwater Rig Drilling Inspection Report," US Department of the Interior website, May 11, 2010, http://www.doi.gov/deepwaterhorizon/upload/05-11-10-MMS-Deepwater-Horizon-Rig-Inspection-Report.pdf (accessed September 16, 2010).

## Chapter 4. Technology Reaching Beyond Our Ability to Control It

1. American Oil and Gas Historical Society, "A History of ROVs," Petroleum History Resources, sites.google.com/site/petroleumhistoryresources/Home/swimming-wrenches (accessed September 13, 2010).

2. Titanic Inquiry Project, "British Wreck Commissioner's Inquiry Report (Report to the British Inquiry Board of Trade Administration, 1912)," www.titanicinquiry.org/BOTInq/BOTReport/BOTRepBOT.php (accessed September 13, 2010).

3. *New York Times* archive, "Col. Astor and Bride, Isidor Straus and Wife, and Maj. Butt Aboard.; RULE OF SEA FOLLOWED," *New York Times*, April 16, 1912, http://query.nytimes.com/mem/archive-free/pdf?res=F70C14FE3E5813738DDDAF0994DC405B828DF1D3 (accessed September 16, 2010).

4. National Aeronautic and Space Administration, "Apollo: History," NASA

website, www.nasa.gov/mission_pages/apollo/missions/apollo13.html (accessed September 13, 2010).

5. David R. Williams, "The Apollo 13 Incident," National Space Science Data Center, nssdc.gsfc.nasa.gov/planetary/lunar/ap13acc.html (accessed September 13, 2010).

6. Apollo 13 Review Board, "Report of Apollo 13 Review Board," June 15, 1970, NASA Technical Report Server, ntrs.nasa.gov/archive/nasa/casi.ntrs.nasa .gov/19700076776_1970076776.pdf (accessed September 13, 2010).

7. Encyclopedia Astronautica, "1970.06.13—Apollo 13 Review Board Publishes Result of Investigation," Encyclopedia Astronautica online, www.astronautix .com/details/apo27567.htm (accessed September 13, 2010).

8. John Waggoner, "Thunder Horse: First of a Generation in the GoM," *Offshore Magazine* (December 1, 2009), www.offshore-mag.com/index/article -display/7929620548/articles/offshore/volume-69/issue-12/top-5_projects/ thunder-horse__first.html (accessed September 13, 2010).

9. BP Press Office, "BP Retires Crazy Horse Name—Project Re-Named Thunder Horse," BP website, February 20, 2002, www.bp.com/genericarticle.do?category Id=2012968&contentId=2000913 (accessed September 13, 2010).

10. Bill Kirton et al., "Thunder Horse Drilling-Riser Break—The Road to Recovery," *Society of Petroleum Engineers Journal of Petroleum Technology* 57, no. 4 (April 2004), www.spe.org/jpt/print/archives/2005/04/JPT2005_04_ODC_focus .pdf.

11. Ray Tyson, "BP's Thunder Horse Oil Field Goes Online," *Petroleum News* 13, no. 25 (June 22, 2008), www.petroleumnews.com/pntruncate/224091963.shtml (accessed September 13, 2010).

12. Glenn Woltman and Dr. Charles Smith, "Investigation of Riser Disconnect and Spill Green Canyon Block, OCS-G 21810, July 5, 2005 (Report for the US Department of the Interior Minerals Management Service Gulf of Mexico OCS Regional Office in New Orleans, August 2006)," www.gomr.mms.gov/ PDFs/2006/2006-058.pdf (accessed September 13, 2010).

13. David Dykes et al., "Investigation of Riser Disconnect and Blowout Mississippi Canyon Block 538 OCS-G 16614 Well #2 (Report for the US Department of the Interior Minerals Management Service Gulf of Mexico OCS Regional Office, February 28, 2000)," www.gomr.mms.gov/PDFs/2001/2001-005.pdf (accessed September 13, 2010).

14. US Department of the Interior Minerals Management Service, "Accident Investigation Report," December 2, 2007, US MMS website, https://www.mms .gov/homepg/offshore/safety/acc_repo/2007/071202.pdf (accessed September 16, 2010).

## Chapter 5. A Rogue from the Start: Influencing World Politics, Gambling with Safety

1. *New York Times* Staff, "BP plc.," *New York Times* online companies index, topics.nytimes.com/top/news/business/companies/bp_plc/index.html (accessed September 15, 2010).

2. *Fortune*, "Global 500 2010: Annual Ranking of the World's Biggest Companies from *Fortune* Magazine," CNN Money website, money.cnn.com/magazines/ fortune/global500/2010/full_list/index.html (accessed September 15, 2010).

3. BP, "BP at a Glance," BP website, www.bp.com/sectiongenericarticle.do?category Id=3&contentId=2006926 (accessed September 15, 2010).

4. BP, "History of BP: The BP Story, from First Oil to Tomorrow's Energies," BP website, http://www.bp.com/multipleimagesection.do?categoryId=2010123& contentId=7059226 (accessed September 16, 2010).

5. Stephen Kinzer, *All the Shah's Men* (Hoboken, NJ: John Wiley & Sons, 2008).

6. *New York Times* Staff, "BP plc.," *New York Times* online companies index, topics.nytimes.com/top/news/business/companies/bp_plc/index.html (accessed September 15, 2010).

7. Ishaan Thardoor, "A Brief History of BP," *Time*, June 2, 2010, http://www.time.com/ time/business/article/0,8599,1993361,00.html (accessed September 16, 2010).

8. John Stauber, "Endangered Wildlife Friends Are Here!" *PR Watch* 8, no. 3 (2001), www.prwatch.org/prwissues/2001Q3/endangered.html (accessed September 13, 2010).

9. Darcy Frey, "How Green Is BP?" *New York Times*, www.nytimes.com/2002/12/08/ magazine/how-green-is-bp.html (accessed September 15, 2010).

10. US Chemical Safety Board, "CSB Investigation of BP Texas City Refinery Disaster Continues as Organizational Issues Are Probed" (October 30, 2006), US Chemical Safety Board website, 76.227.217.14/newsroom/detail .aspx?nid=215 (accessed September 13, 2010).

11. Guy Chazan, "BP Fined $87 Million Over Explosion," *Wall Street Journal*, November 2, 2009, online.wsj.com/article/SB10001424052748703392004574505034081842414.html (accessed September 15, 2010).

12. *BBC News*, "BP Chief Executive Browne Resigns," *BBC News* website, May 1, 2007, news.bbc.co.uk/2/hi/business/6612703.stm (accessed September 15, 2010).

13. Katie Paul, "Behind Petroleum," *Newsweek*, June 10, 2010, www.newsweek .com/2010/06/10/behind-petroleum.html (accessed September 13, 2010).

## Chapter 6. How Did We Get Here? A Brief History of Offshore Drilling

1. American Oil and Gas Historical Society, "Offshore Oil History," Petroleum History Resources, webcache.googleusercontent.com/search?q=cache:BMVBdQ Rp920J:sites.google.com/site/petroleumhistoryresources/Home/offshore-oil -history+kermac+16&cd=2&hl=en&ct=clnk&gl=us&client=safari (accessed September 13, 2010).

2. American Oil and Gas Historical Society, "Ohio Offshore Wells" (January 26, 2010), Petroleum History Resources, sites.google.com/site/ petroleumhistoryresources/Home/offshore-oil-history/offshore-wells-in-ohio (accessed September 13, 2010).

3. American Oil and Gas Historical Society, "Offshore Oil History," Petroleum History Resources, webcache.googleusercontent.com/search?q=cache:BMVBdQ Rp920J:sites.google.com/site/petroleumhistoryresources/Home/offshore-oil -history+kermac+16&cd=2&hl=en&ct=clnk&gl=us&client=safari (accessed September 13, 2010).

4. Penn Energy, "Pending Merger of Giant Drillers Signals Long Life for Deepwater Operations," Penn Energy website, www.pennenergy.com/index/petroleum/ display/307363/articles/offshore/volume-67/issue-9/supplement/pending-merger -of-giant-drillers-signals-long-life-for-deepwater-operations.html (accessed September 13, 2010).

5. International Petroleum Museum and Exposition, "The Mr. Charlie: The Details," International Petroleum Museum and Exposition website, www .rigmuseum.com/charlie2.html (accessed September 10, 2010).

6. Offshore Energy Center, "2000 Technology Pioneers: Drilling (Mobile Drilling Units) First and Second Generation Semisubmersible Drilling Rigs," Offshore Energy Center website, http://www.oceanstaroec.com/fame/2000/ semisubmersible.htm (accessed September 16, 2010).

7. Transocean, "Discoverer Enterprise," Transocean website, www.deepwater.com/ fw/main/Discoverer-Enterprise-141.html (accessed September 14, 2010).

8. Bureau of Ocean Energy Management, Regulation and Enforcement, "OCS Lands Act History," BOEMRE website, www.boemre.gov/aboutmms/OCSLA/ ocslahistory.htm (accessed September 13, 2010).

9. National Ocean Industries Association, "History of Offshore," NOIA website, www.noia.org/website/article.asp?id=123 (accessed September 13, 2010).

10. The United Nations Law of the Sea Research Center, "Law of the Sea Treaty," National Center for Public Policy Research, 2010, http://www .unlawoftheseatreaty.org/ (accessed September 16, 2010).

11. US Environmental Protection Agency, "Oil Pollution Act Overview," US Environmental Protection Agency website, http://www.epa.gov/oem/content/ lawsregs/opaover.htm (accessed September 16, 2010).

12. US Energy Information Administration, "Moratorium on Offshore Drilling (1990)," US Energy Information Administration website, http://www.eia.doe .gov/oil_gas/natural_gas/analysis_publications/ngmajorleg/moratorium.html (accessed September 16, 2010).

13. Greenpeace, "Offshore Disaster: Timeline of Offshore Drilling, Spills, and Regulations," US Climate Network site, 2010, http://www.usclimatenetwork .org/resource-database/offshore-disaster-timeline-of-offshore-oil-drilling-spills -and-regulation (accessed September 17, 2010).

14. William A. Trapmann, "Chapter 4: Offshore Development and Production," in *Natural Gas 1998: Issues and Trends,* edited by James Tobin, US Energy Information Administration, April 1999, www.eia.doe.gov/pub/oil_gas/natural_ gas/analysis_publications/natural_gas_1998_issues_trends/pdf/chapter4.pdf (accessed September 13, 2010).

15. Bureau of Ocean Energy Management, Regulation and Enforcement, "OCS Lands Act History," BOEMRE website, www.boemre.gov/aboutmms/OCSLA/ ocslahistory.htm (accessed September 13, 2010).

16. Clifford Krauss, "BP Finds Giant Oil Field Deep in the Gulf of Mexico," *New York Times,* September 2, 2010, http://www.nytimes.com/2009/09/03/business/ global/03oil.html (accessed September 16, 2010).

17. Dan Eggen and Steven Muffson, "Bush Rescind's Father's Offshore Oil Ban," *Washington Post,* July 15, 2008, http://www.washingtonpost.com/wp-dyn/content/ article/2008/07/14/AR2008071401049.html (accessed September 16, 2010).

18. John M. Broder, "Obama to Open Offshore Areas to Oil Drilling for the First Time," *New York Times,* March 31, 2010, http://www.nytimes.com/2010/03/31/ science/earth/31energy.html (accessed September 16, 2010).

19. US Energy Information Administration Independent Statistics and Analysis, "Gulf of Mexico Fact Sheet," September 3, 2010, www.eia.doe.gov/special/ gulf_of_mexico/index.cfm (accessed September 13, 2010).

20. Brett Clanton, "Deep-Water Ban Sending Ripples Through Houston," *Houston Chronicle,* June 6, 2010, www.chron.com/disp/story.mpl/business/deepwaterhorizon/7039523.html (accessed September 13, 2010).

## Chapter 7. Corporate Shape-Shifting and the BP–Government Merger

1. BP Press Centre, "BP Establishes $20 Billion Claims Fund for Deepwater Horizon Spill and Outlines Dividend Decisions," BP website, June 16, 2010, www.bp.com/genericarticle.do?categoryId+2012968&contld=7062966 (accessed September 14, 2010).

2. Deepwater Horizon Joint Investigation Committee, "US Coast Guard Hearing on the Deepwater Horizon Explosion from Houston, Day 1," testimony of Daun Winslow, August 23, 2010, www.c-span.org/Watch/Media/2010/08/23/HP/A/37295/US+Coast+Guard+Hearing+on+the+Deepwater+Horizon+Explosion+from+Houston+Day+1.aspx (accessed September 14, 2010).

3. Claire Suddath, "A Brief History of the War on Drugs," *Time*, March 25, 2009, www.time.com/time/world/article/0,8599,1887488,00.html (accessed September 14, 2010).

4. Lisa Stark, "Coast Guard Chief: Age of Fleet 'Putting Our Crews at Risk,'" *ABC News*, February 12, 2010, blogs.abcnews.com/thenote/2010/02/coast-guard-chief-age-of-fleet-putting-our-crews-at-risk.html (accessed September 14, 2010).

5. Deepwater Horizon Joint Investigation Committee, "US Coast Guard Hearing on the Deepwater Horizon Explosion from Houston, Day 1," testimony of Daun Winslow, August 23, 2010, www.c-span.org/Watch/Media/2010/08/23/HP/A/37295/US+Coast+Guard+Hearing+on+the+Deepwater+Horizon+Explosion+from+Houston+Day+1.aspx (accessed September 14, 2010).

6. Aaron Solomon and John Mehta, "Haphazard Firefighting Might Have Sunk BP Oil Rig," Center for Public Integrity website, July 27, 2010, www.publicintegrity.org/articles/entry/2286 (accessed September 14, 2010).

7. BP Press Centre, "BP Offers Full Support to Transocean After Drilling Rig Fire," BP website, April 21, 2010, www.bp.com/genericarticle.do?categoryId=2012968&contentId=7061458 (accessed September 14, 2010).

8. White House, "Remarks by the President at Earth Day Reception," White House Office of the Press Secretary website, April 22, 2010, www.whitehouse.gov/the-press-office/remarks-president-earth-day-reception (accessed September 14, 2010).

9. White House, "Statement by the Press Secretary on the President's Oval Office Meeting to Discuss the Situation in the Gulf of Mexico," White House Office of the Press Secretary website, April 22, 2010, www.whitehouse.gov/the-press-office/statement-press-secretary-presidents-oval-office-meeting-discuss-situation-gulf-mex (accessed September 14, 2010).

10. Deepwater Horizon Response, "UPDATE 5: Search and Rescue Continues; Joint Environmental Response Efforts in Place," Deepwater Horizon Unified Command official website, April 23, 2010, www.deepwaterhorizonresponse.com/go/doc/2931/529003 (accessed September 14, 2010).

11. Deepwater Horizon Joint Investigation, "Investigation of Deepwater Horizon Explosion, Harry Thierens Testimony Part 1, August 25, 2010," C-Span Video Library, www.c-spanvideo.org/program/295172-2 (accessed September 10, 2010).

12. Deepwater Horizon Response, "UPDATE 7: Unified Command Continues

to Respond to Deepwater Horizon," Deepwater Horizon Unified Command official website, April 25, 2010, www.deepwaterhorizonresponse.com/go/doc/2931/529883 (accessed September 14, 2010).

13. Campbell Robertson and Leslie Kaufman, "Size of Spill in Gulf of Mexico Is Larger Than Thought," *New York Times*, April 28, 2010, www.nytimes.com/2010/04/29/us/29spill.html (accessed September 14, 2010).

14. Huffington Post, "Oil Execs Called to Testify Before Congress on Gulf Coast Spill, Consumer Pricing," Huffington Post website, April 29, 2010, www.huffingtonpost.com/2010/04/29/oil-execs-called-to-testi_n_557126.html (accessed September 14, 2010).

15. White House, "Press Briefing on the BP Oil Spill in the Gulf Coast," White House Office of the Press Secretary website, April 29, 2010, www.whitehouse.gov/the-press-office/press-briefing-bp-oil-spill-gulf-coast (accessed September 14, 2010).

16. *New Orleans Times-Picayune* blog, "Gov. Jindal Issues Emergency Declaration in Oil Leak in Gulf of Mexico," April 29, 2010, http://www.nola.com/news/index.ssf/2010/04/gov_jindal_issues_emergency_de.html (accessed September 16, 2010).

17. Stephanie Condon, "Axelrod: No New Drilling Until Cause of Oil Spill Determined," *CBS News*, April 30, 2010, www.cbsnews.com/8301-503544_162-20003846-503544.html (accessed September 14, 2010).

18. Jake Sherman and Meredith Shiner, "Democrats: BP Answers Insufficient," *Politico*, May 4, 2010, www.politico.com/news/stories/0510/36777.html (accessed September 14, 2010).

19. NPR, "BP Will Pay for Gulf Oil Spill Disaster, CEO Says," NPR website, May 3, 2010, www.npr.org/templates/story/story.php?storyId=126468782 (accessed September 14, 2010).

20. Daily Telegraph Staff, "BP's Tony Hayward: The Gaffes," *Daily Telegraph*, July 26, 2010, www.telegraph.co.uk/finance/newsbysector/energy/oilandgas/7910167/BPs-Tony-Hayward-the-gaffes.html (accessed September 14, 2010).

21. Clifford Krauss, Henry Fountain, and John M. Broder, "Acrimony Behind the Scenes of Gulf Oil Spill," *New York Times*, August 26, 2010, www.nytimes.com/2010/08/27/us/27well.html?_r=4&pagewanted=1&hp (accessed September 14, 2010).

22. Jim Efstathiou Jr., "BP, Halliburton, Transocean Blame Each Other in Gulf Oil Spill," *Bloomberg Businessweek*, May 10, 2010, www.businessweek.com/news/2010-05-10/bp-halliburton-transocean-blame-each-other-in-gulf-oil-spill.html (accessed September 14, 2010).

23. Jennifer Dlouhy, "Pressure Readings Prior to Gulf Spill Eyed," *San Antonio Express-News*, May 5, 2010, www.mysanantonio.com/news/Pressure_readings_prior_to_Gulf_spill_eyed_93553019.html (accessed September 14, 2010).

24. Steven Mufson and David Fahrenholdt, "Oil Spill Investigators Find Critical Problems in Blowout Preventer," *Washington Post*, May 13, 2010, www.washingtonpost.com/wp-dyn/content/article/2010/05/12/AR2010051202190.html (accessed September 14, 2010).

25. David S. Hilzenrath, "BP Executive Says Blowout Preventer Was Not Connected Properly," *Washington Post*, August 25, 2010, www.washingtonpost.com/wp-dyn/content/article/2010/08/25/AR2010082504173.html (accessed September 14, 2010).

26. Ben Geman, "Coast Guard, MMS to Formally Launch Joint Investigation into Gulf Oil Spill," The Hill, May 8, 2010, thehill.com/blogs/e2-wire/677-e2 -wire/96805-coast-guard-mms-to-formally-launch-joint-investigation-into-gulf -oil-spill (accessed September 14, 2010).

27. Anna Driver, "BP Capturing 5,000 bpd at Gulf Leak Site," Reuters AlertNet, May 20, 2010, www.alertnet.org/thenews/newsdesk/N20148849.htm (accessed September 14, 2010).

28. Jaquetta White, "Oil Leak Measurements Skewed by Natural Gas in the Mix, BP Says," New Orleans Times-Picayune, May 21, 2010, www.nola.com/news/ gulf-oil-spill/index.ssf/2010/05/oil_leak_measurements_skewed_b.html (accessed September 14, 2010).

29. Eljefebob, "BP Announces Success of Riser Insertion Tool—But Remains Coy," Daily Hurricane, May 17, 2010, dailyhurricane.com/2010/05/bp-announces- success-of-riser-insertion-tool---but-remains-coy.html (accessed September 14, 2010).

30. Eljefebob, "Two New Subsea Views of BP Blowout. It Ain't Pretty," Daily Hurricane, May 19, 2010, dailyhurricane.com/2010/05/two-new-subsea-views-of -bp-blowout-it-aint-pretty.html (accessed September 14, 2010).

31. Bettina Boxall, "Gulf Oil Leak Rate Much Higher Than Reported, Professor Says," Los Angeles Times, May 14, 2010, articles.latimes.com/2010/may/14/ nation/la-na-oil-spill-measure-20100514 (accessed September 14, 2010).

32. Ben Geman, "BP Exec Disputes Higher Spill Rate Estimates," The Hill, May 14, 2010, thehill.com/blogs/e2-wire/677-e2-wire/97955-bp-exec-disputes-higher -spill-rate-estimates (accessed September 14, 2010).

33. Douglas J. Suttles, "Re: Source Control Subsea Dispersant Plan (Letter to Rear Admiral James A. Watson, July 6, 2010)," Deepwater Horizon Unified Command website, www.deepwaterhorizonresponse.com/go/doc/2931/780031 (accessed September 14, 2010).

34. Douglas J. Suttles, "Exemption to Dispersant Monitoring and Assessment Directive—Addendum 3 (Letter to Rear Admiral James A. Watson, July 6, 2010)," Deepwater Horizon Unified Command website, www.deepwaterhorizonresponse.com/external/content/ document/2931/780047/1/071110.PDF (accessed September 14, 2010).

35. White House, "Remarks by the President on the Ongoing Oil Spill Response," White House Office of the Press Secretary website, May 14, 2010, www .whitehouse.gov/the-press-office/remarks-president-ongoing-oil-spill-response (accessed September 14, 2010).

36. White House, "Remarks by the President on the Gulf Oil Spill," White House Office of the Press Secretary website, May 27, 2010, www.whitehouse.gov/ the-press-office/remarks-president-gulf-oil-spill (accessed September 14, 2010).

37. Judith Evans, "Boris Johnson Attacks America's 'Anti-British Rhetoric' on BP," Times Online, June 10, 2010, www.timesonline.co.uk/tol/news/uk/ article7147278.ece (accessed September 14, 2010).

38. Jane Wardell, "BP's Stock Plunges to 18-Year Low as Company's Survival Could Hang in the Balance," Huffington Post website, June 1, 2010, www .huffingtonpost.com/2010/06/01/bp-shares-plunge-to-18yea_n_595796.html (accessed September 14, 2010).

39. Huffington Post, "Ed Markey: BP 'Lying or Incompetent' About Scope of Gulf Oil Spill," Huffington Post website, May 30, 2010, www.huffingtonpost

.com/2010/05/30/ed-markey-bp-lying-or-inc_n_594800.html (accessed
September 14, 2010).

40. Andrew Ross Sorkin, "Imagining the Worst in BP's Future," *New York Times*,
June 7, 2010, www.nytimes.com/2010/06/08/business/08sorkin.html (accessed
September 14, 2010).

41. BP Press Centre, "BP Establishes $20 Billion Claims Fund for Deepwater
Horizon Oil Spill and Outlines Dividend Decisions," BP website, June 16,
2010, www.bp.com/genericarticle.do?categoryId=2012968&contentId=7062966
(accessed September 15, 2010).

42. Ann Gerhart, "BP Chairman Talks About the 'Small People,' Further Angering
Gulf," *Washington Post*, June 17, 2010, www.washingtonpost.com/wp-dyn/
content/article/2010/06/16/AR2010061605528.html (accessed September 14,
2010).

43. Deepwater Horizon Incident Joint Information Center, "Transcript—
Press Briefing by National Incident Commander Admiral Thad Allen,"
Deepwater Horizon Unified Command website, August 11, 2010, www
.deepwaterhorizonresponse.com/go/doc/2931/857771 (accessed September 14,
2010).

44. Joel Achenbach, "With BP's Know-How and US Authority, the Macondo Well
Was Plugged," *Washington Post*, August 21, 2010, www.washingtonpost.com/
wp-dyn/content/article/2010/08/20/AR2010082005747.html?hpid=topnews&sid
=ST2010082005804 (accessed September 14, 2010).

45. US Department of Energy, "Deepwater Horizon Response: Department of
Energy Actions on BP Oil Spill," US Department of Energy website, www
.energy.gov/open/oil_spill_updates.htm (accessed September 14, 2010).

46. Deepwater Horizon Incident Joint Information Center, "Transcript—Press Brief
with National Incident Commander Admiral Thad Allen," Deepwater Horizon
Unified Command website, July 16, 2010, www.deepwaterhorizonresponse.com/
go/doc/2931/789331 (accessed September 14, 2010).

## Chapter 8. How Not to Save an Ecosystem: Dispersants, Skimming, and Booming

1. US Environmental Protection Agency and the US Coast Guard, "Dispersant
Monitoring and Assessment Directive for Subsurface Dispersant Application
(Directive to BP, May 10, 2010)," http://www.epa.gov/bpspill/dispersants/
subsurface-dispersant-directive-final.pdf (accessed September 16, 2010).

2. Douglas J. Suttles, "Re: May 19, 2010 Addendum 2 to Dispersant Monitoring
and Assessment Directive (Letter to Mary Landry and Samuel Coleman, May
20, 2010)," http://www.epa.gov/bpspill/dispersants/5-21bp-response.pdf (accessed
September 16, 2010).

3. US Environmental Protection Agency and the US Coast Guard, "Dispersant
Monitoring and Assessment Directive for Subsurface Dispersant Application—
Addendum 3 (Directive to BP, May 26, 2010)," http://www.epa.gov/bpspill/
dispersants/directive-addendum3.pdf (accessed September 16, 2010).

4. David A. Fahrenhold and Steven Mufson, "Documents Indicate Heavy Use of
Dispersants in Gulf Oil Spill," *Washington Post*, August 1, 2010, http://www
.washingtonpost.com/wpdyn/content/article/2010/07/31/AR2010073102381.
html (accessed September 17, 2010).

5. Jane Lubchenco et al., "Deepwater Horizon/BP Oil Budget: What Happened to All the Oil? (study by National Incident Command)," Deepwater Horizon Official Response website, August 1, 2010, http://www .deepwaterhorizonresponse.com/posted/2931/Oil_Budget_description_8_3_ FINAL.844091.pdf (accessed September 17, 2010).

6. David Biello, "Slick Solution: How Microbes Will Clean Up the Deepwater Horizon Oil Spill," Scientific American, May 25, 2010, www.scientificamerican.com/ article.cfm?id=how-microbes-clean-up-oil-spills (accessed September 14, 2010).

7. Justin Gillis, "US Finds Most Oil from Spill Poses Little Additional Risk," New York Times, August 8, 2010, www.nytimes.com/2010/08/04/science/earth/04oil .html?_r=2&ref=earth (accessed September 14, 2010).

8. Emily Gertz, "Marine Toxicologist Susan Shaw Dives into Gulf Spill, Talks Dispersants and Food Web Damage," Onearth blog, June 24, 2010, www .onearth.org/node/2253 (accessed September 14, 2010).

9. US Environmental Protection Agency, "Whitman Details Ongoing Agency Efforts to Monitor Disaster Sites, Contribute to Cleanup Efforts," EPA website, September 18, 2001, www.epa.gov/wtc/stories/headline_091801.htm (accessed September 14, 2010).

10. Marian Wang, "In Gulf Spill, BP Using Dispersants Banned in UK," ProPublica blog, May 18, 2010, www.propublica.org/blog/item/In-Gulf-Spill-BP-Using -Dispersants-Banned-in-UK (accessed September 14, 2010).

11. Susan Shaw, "What the EPA Dispersant Tests Fail to Tell Us," Marine Environmental Research Institute website, www.meriresearch.org/Portals/0/ Documents/Susan%20Shaw%20Statement%20Re%20EPA%20Report%20 August%2013.pdf (accessed September 14, 2010).

12. Emily Gertz, "Marine Toxicologist Susan Shaw Dives into Gulf Spill, Talks Dispersants and Food Web Damage," Onearth blog, June 24, 2010, www .onearth.org/node/2253 (accessed September 14, 2010).

13. Michael J. Hemmer et al., "Comparative Toxicity of Eight Oil Dispersant Products on Two Gulf of Mexico Aquatic Test Species," US Environmental Protection Agency Office of Research and Development study, June 30, 2010, http://www.epa.gov/bpspill/reports/ComparativeToxTest.Final.6.30.10.pdf (accessed September 17, 2010).

14. Suzanne Goldenberg, "BP Oil Spill: Obama Administration's Scientists Admit Alarm over Chemicals," Guardian, August 3, 2010, http://www.guardian.co.uk/ environment/2010/aug/03/gulf-oil-spill-chemicals-epa (accessed September 17, 2010).

15. E. Eric Adams and Scott A. Socolofsky, "Review of Deep Oil Spill Modeling Activity Supported by the DeepSpill JIP and Offshore Operators Committee," BOEMRE website, December 2004, revised February 2005, www.boemre.gov/ tarprojects/377/Adams%20Review%204.pdf (accessed September 14, 2010).

16. Erin Hartness, "UNC Researchers Say Much of Oil Spill Floating Under Gulf," WRAL.com, May 20, 2010, www.wral.com/news/local/story/7640522 (accessed September 14, 2010).

17. Vickie Chachere, "USF Scientists Detect Oil on Gulf Floor," University of South Florida News, August 17, 2010, usfweb3.usf.edu/absoluteNM/ templates/?a=2604&z=120 (accessed September 14, 2010).

18. Justin Gillis, "Plume of Oil Below Surface Raise New Concerns," New York Times, June 8, 2010, http://www.nytimes.com/2010/06/09/us/09spill.html (accessed September 17, 2010).

19. Danielle Thomas, "Gulf Fishermen Say They're Not Convinced Seafood Is Safe," WLOX.com, August 4, 2010, www.wlox.com/Global/story .asp?S=12927774 (accessed September 14, 2010).

20. Bob Warren, "State Authorities Say Fish Kill in St. Bernard Parish Waters Likely Caused by Low Oxygen Levels," *New Orleans Times-Picayune*, August 23, 2010, www.nola.com/news/gulf-oil-spill/index.ssf/2010/08/large_fish_kill_ reported_in_st.html (accessed September 14, 2010).

21. Suzanne Goldenberg, "BP Oil Spill: Obama Administration's Scientists Admit Alarm over Chemicals," *Guardian*, August 3, 2010, http://www.guardian.co.uk/ environment/2010/aug/03/gulf-oil-spill-chemicals-epa (accessed September 17, 2010).

22. Susan Shaw, "What the EPA Dispersant Tests Fail to Tell Us," Marine Environmental Research Institute website, www.meriresearch.org/Portals/0/ Documents/Susan%20Shaw%20Statement%20Re%20EPA%20Report%20 August%2013.pdf (accessed September 14, 2010).

23. Jane Lubchenco et al., "Deepwater Horizon/BP Oil Budget: What Happened to All the Oil? (study by National Incident Command)," Deepwater Horizon Official Response website, August 1, 2010, http://www .deepwaterhorizonresponse.com/posted/2931/Oil_Budget_description_8_3_ FINAL.844091.pdf (accessed September 17, 2010).

24. The White House Office of the Press Secretary, "Press Briefing by Press Secretary Robert Gibbs, Admiral Thad Allen, Carol Browner, and Dr. Lubchenco, 8/4/2010," The White House, http://www.whitehouse.gov/the-press-office/press -briefing-press-secretary-robert-gibbs-admiral-thad-allen-carol-browner-and-dr (accessed September 17, 2010).

25. Justin Gillis, "US Finds Most Oil from Spill Poses Little Additional Risk," *New York Times*, August 8, 2010, www.nytimes.com/2010/08/04/science/earth/04oil .html?_r=2&ref=earth (accessed September 14, 2010).

26. Robert Lee Hotz, "Much Oil Remains in Gulf, Researchers Estimate," *Wall Street Journal*, August 17, 2010, http://online.wsj.com/article/SB100014240527487048 68604575434074237252604.html (accessed September 17, 2010).

27. Elizabeth Shogren, "Scientists Split on Gulf Oil Estimates," *NPR*, August 19, 2010, http://www.npr.org/templates/story/story.php?storyId=129306358 (accessed September 17, 2010).

28. Kate Sheppard, "Return of the BP Cover-Up," *Mother Jones*, August 19, 2010, motherjones.com/blue-marble/2010/08/noaa-no-data-spill-claim-least-two -months (accessed September 14, 2010).

29. Dan Froomkin, "NOAA Claims Scientists Reviewed Controversial Report; The Scientists Say Otherwise," Huffington Post website, August 20, 2010, http:// www.huffingtonpost.com/2010/08/20/noaa-claims-scientists-re_n_689428.html (accessed September 17, 2010).

30. Craig Pittman, "USF Scientists Find Oil Spill Damage to Critical Marine Life," *St. Petersburg Times*, August 18, 2010, http://www.tampabay.com/news/ education/college/usf-scientists-find-oil-spill-damage-to-critical-marine -life/1115706 (accessed September 17, 2010).

31. Justin Gillis and John Collins Rudolf, "Gulf Oil Spill Is Not Breaking Down So Fast, Study Says," *New York Times*, August 19, 2010, http://www.nytimes .com/2010/08/20/science/earth/20plume.html (accessed September 17, 2010).

32. Lynn Yarris, "Study Shows Deepwater Oil Plume in Gulf Degraded by Microbes,"

Lawrence Berkeley National Laboratory News Center website, August 24, 2010, newscenter.lbl.gov/news-releases/2010/08/24/deepwater-oil-plume-microbes (accessed September 14, 2010).

33. Jane Tierney, "Berkeley Scientist Advises on Clean-up in Gulf Oil Spill," berkeleyside.com, June 4, 2010, http://www.berkeleyside.com/2010/06/04/berkeley-scientist-advises-on-clean-up-in-gulf-oil-spill/ (accessed September 17, 2010).

34. Energy Biosciences Institute, "EBI at a Glance," Energy Biosciences Institute website, http://www.energybiosciencesinstitute.org/index.php?option=com_content&task=blogsection&id=2&Itemid=3 (accessed September 17, 2010).

35. Associated Press wires, "7 Gulf Oil Spill Cleanup Workers Hospitalized," Huffington Post website, May 27, 2010, http://www.huffingtonpost.com/huff-wires/20100527/us-gulf-oil-spill-sick-workers/ (accessed September 17, 2010).

36. Riki Ott, "At What Cost? BP Spill Responders Told to Forgo Precautionary Health Measures in Cleanup," Huffington Post website, May 17, 2010, http://www.huffingtonpost.com/riki-ott/at-what-cost-bp-spill-res_b_578784.html (accessed September 17, 2010).

37. Melanie Trottman, "OSHA Says BP Workers Don't Need Respirators," *Wall Street Journal*, June 4, 2010, http://online.wsj.com/article/SB10001424052748704764404575286180491707288.html (accessed September 17, 2010).

38. James C. McKinley Jr. and James Collins Rudolf, "Experts Express Doubts on Sand Berm Proposal," *New York Times*, May 21, 2010, http://www.nytimes.com/2010/05/22/us/22berms.html (accessed September 17, 2010).

39. Julie Cart, "Gulf Oil Spill: Barrier Island Berm Plan Runs Aground," *Los Angeles Times* Greenspace blog, June 23, 2010, http://latimesblogs.latimes.com/greenspace/2010/06/gulf-oil-spill-barrier-island-berm-plan-runs-aground.html (accessed September 17, 2010).

40. Mark Schleifstein, "Sand Berm Defense Against Oil from Gulf of Mexico Spill Gets $60 Million Financing Installment," *New Orleans Times-Picayune*, August 19, 2010, http://www.nola.com/news/gulf-oil-spill/index.ssf/2010/08/sand_berm_defense_against_oil.html (accessed September 17, 2010).

41. Associated Press news wire, "'A Whale' Too Big to Clean Gulf Oil Spill," Huffington Post website, July 17, 2010, http://www.huffingtonpost.com/2010/07/18/a-whale-too-big-to-clean_n_650274.html (accessed September 17, 2010).

42. Adam Gabbatt, "BP Oil Spill: Kevin Costner's Oil-Water Separation Machines Help with Clean-up," *Guardian*, June 16, 2010, http://www.guardian.co.uk/environment/2010/jun/16/kevin-costner-oil-spill-machines (accessed September 17, 2010).

## Chapter 9. Top Cap, Top Hat, Top Kill, Capping Stack: Making It Up as We Go Along

1. Bettina Boxall, "Gulf Oil Leak Rate Much Higher Than Reported, Professor Says," *Los Angeles Times*, May 14, 2010, articles.latimes.com/2010/may/14/nation/la-na-oil-spill-measure-20100514 (accessed September 14, 2010).

2. US Department of the Interior, "Admiral Allen, Dr. McNutt Provide Updates on Progress of Scientific Teams Analyzing Flow Rates from BP's Well," US Department of the Interior website, June 10, 2010, www.doi.gov/news/

pressreleases/Admiral-Allen-Dr-McNutt-Provide-Updates-on-Progress-of-Scientific-Teams-Analyzing-Flow-Rates-from-BPs-Well.cfm (accessed September 14, 2010).

3. Joe Flaherty and Mary Pat Stephens, "Oil Industry Cleanup Organization Swamped by BP Spill," *Washington Post*, June 29, 2010, www.washingtonpost.com/wp-dyn/content/article/2010/06/29/AR2010062905384.html?sid=ST2010062905387 (accessed September 14, 2010).

4. Deepwater Horizon Joint Investigation, "Investigation of Deepwater Horizon Explosion, Harry Thieren's Testimony Part 1, August 25, 2010," C-Span Video Library, www.c-spanvideo.org/program/295172-2 (accessed September 10, 2010).

5. George Talbot, "UPDATE—BP Refutes Executive Who Said Company 'Cut the Flow' of Oil from Damaged Rig," *Mobile Press-Register*, May 3, 2010, blog.al.com/live/2010/05/bp_official_weve_significantly.html (accessed September 14, 2010).

6. Transocean, "Discoverer Enterprise," Transocean website, www.deepwater.com/fw/main/Discoverer-Enterprise-141.html (accessed September 14, 2010).

7. Rebecca Mowbray, "Gulf of Mexico Oil Spill Containment Box Comes with No Guarantees," *New Orleans Times-Picayune*, May 7, 2010, www.nola.com/news/gulf-oil-spill/index.ssf/2010/05/gulf_of_mexico_oil_spill_conta.html (accessed September 15, 2010).

8. David Wethe, "BP Oil-Collection Chamber Clogs, Removed from Well," *Bloomberg Businessweek*, May 9, 2010, www.businessweek.com/news/2010-05-09/bp-oil-collection-chamber-clogs-removed-from-well-update1-.html (accessed September 14, 2010).

9. Jaquetta White, "Smaller 'Top Hat' Containment Box Being Lowered over Gulf Oil Leak," *New Orleans Times-Picayune*, May 11, 2010, http://www.nola.com/news/gulf-oil-spill/index.ssf/2010/05/smaller_top_hat_containment_bo.html (accessed September 17, 2010).

10. Jaquetta White, "Oil Leak Measurements Skewed by Natural Gas in the Mix, BP Says," *New Orleans Times-Picayune*, May 21, 2010, www.nola.com/news/gulf-oil-spill/index.ssf/2010/05/oil_leak_measurements_skewed_b.html (accessed September 15, 2010).

11. CNN Wire Staff, "After Delay, BP Restarts 'Top Kill' Effort," CNN, May 27, 2010, www.cnn.com/2010/US/05/27/gulf.oil.spill/index.html (accessed September 14, 2010).

12. Jim Tankersley, "'Top Kill' Stops Gulf Oil Leak for Now, Official Says," *Los Angeles Times*, May 28, 2010, articles.latimes.com/2010/may/28/nation/la-na-oil-spill-top-kill-20100528 (accessed September 15, 2010).

13. CNN Wire Staff, "After Delay, BP Restarts 'Top Kill' Effort," CNN, May 27, 2010, www.cnn.com/2010/US/05/27/gulf.oil.spill/index.html (accessed September 14, 2010).

14. David Hammer, "Discovery of Second Pipe in Deepwater Horizon Riser Stirs Debate among Experts," *New Orleans Times-Picayune*, July 9, 2010, http://www.nola.com/news/gulf-oil-spill/index.ssf/2010/07/post_19.html (accessed September 17, 2010).

15. Phaedra Friend Troy, "LMRP Oil Cap Sees Success," *Penn Energy*, June 4, 2010, www.pennenergy.com/index/petroleum/display/7714265166/articles/pennenergy/oil-spill-gulf-of-mexico-2010/lmrp-oil_spill_containment.html (accessed September 15, 2010).

16. Admiral Thad Allen, letter to Bob Dudley, July 8, 2010, Deepwater Horizon

Unified Command website, www.deepwaterhorizonresponse.com/external/
content/document/2931/766283/1/BP%20letter%208%20July%5B1%5D.pdf
(accessed September 14, 2010).

17. Bob Dudley, "Gulf Coast Restoration," *Deepwater Horizon* Unified Command
website, July 9, 2010, http://www.deepwaterhorizonresponse.com/external/
content/document/2931/770167/1/Ltr%20to%20Admiral%20Allen%20
with%20attachments%20-%2009%2007%2010.pdf.

18. Admiral Thad Allen, letter to Bob Dudley, July 8, 2010, Deepwater Horizon
Unified Command website, www.deepwaterhorizonresponse.com/external/
content/document/2931/766283/1/BP%20letter%208%20July%5B1%5D.pdf
(accessed September 14, 2010).

19. Bob Dudley, letter to Admiral Thad Allen, July 9, 2010, Deepwater Horizon
Unified Command website, www.deepwaterhorizonresponse.com/external/
content/document/2931/770167/1/Ltr%20to%20Admiral%20Allen%20
with%20attachments%20-%2009%2007%2010.pdf (accessed September 14,
2010).

20. Kent Wells, transcript of technical briefing conference call moderated
by Marcella Christophe, July 13, 2010, BP website, www.bp.com/
liveassets/bp_internet/globalbp/globalbp_uk_english/incident_response/
STAGING/local_assets/downloads_pdfs/kent_wells_technical_briefing_
transcript_13_07_0730CDT.pdf (accessed September 15, 2010).

21. Deepwater Horizon Response, "Transcript—Press Briefing with National
Incident Commander Admiral Thad Allen," Deepwater Horizon Unified
Command website, July 13, 2010, www.deepwaterhorizonresponse.com/go/
doc/2931/779967 (accessed September 15, 2010).

22. Doug Suttles, conference call moderated by Bill Slavin, July 18, 2010, BP
website, www.bp.com/liveassets/bp_internet/globalbp/globalbp_uk_english/
incident_response/STAGING/local_assets/downloads_pdfs/technical_briefing_
transcript_18_07_0730CDT.pdf (accessed September 14, 2010).

23. Kent Wells, conference call moderated by Jerrin Bodo, July 17, 2010, BP
website, www.bp.com/liveassets/bp_internet/globalbp/globalbp_uk_english/
incident_response/STAGING/local_assets/downloads_pdfs/kent_wells_
technical_briefing_transcript_17_07_0730CDT.pdf (accessed September 14,
2010).

24. Kent Wells, conference call moderated by Daren Beaudo, July 19, 2010, BP
website, http://www.bp.com/liveassets/bp_internet/globalbp/globalbp_uk_
english/incident_response/STAGING/local_assets/downloads_pdfs/BP_
technical_audio_07192010.pdf (accessed September 14, 2010).

25. BP Press Centre, "MC252 Well Reaches Static Condition," BP website,
August 4, 2010, www.bp.com/genericarticle.do?categoryId=2012968&
contentId=7064173 (accessed September 14, 2010).

26. Chris Baltimore, "BP Says Macondo Well Cement Performing as Expected,"
Reuters, August 6, 2010, www.reuters.com/article/idUSN0623738520100806
(accessed September 14, 2010).

27. US Coast Guard, "Press Briefing by National Incident Commander Admiral
Thad Allen Aug. 10, 2010," US Coast Guard Visual Information Gallery, cgvi
.uscg.mil/media/main.php?g2_itemId=968590 (accessed September 14, 2010).

28. Kent Wells, transcript of technical briefing moderated by Scott Dean, August
10, 2010, BP website, www.bp.com/liveassets/bp_internet/globalbp/globalbp_

uk_english/incident_response/STAGING/local_assets/downloads_pdfs/BP_
technical_audio_87627186.pdf (accessed September 15, 2010).

29. Deepwater Horizon Joint Information Center, "Transcript of Thad Allen's
Aug. 16 Briefing on the BP Oil Spill," *McClatchy*, August 16, 2010, www
.mcclatchydc.com/2010/08/16/99254/transcript-of-thad-allens-aug.html
(accessed September 14, 2010).

30. Deepwater Horizon Incident Joint Information Center, "Transcript—
Press Briefing by National Incident Commander Admiral Thad Allen,"
Deepwater Horizon Unified Command website, August 27, 2010, www
.deepwaterhorizonresponse.com/go/doc/2931/886987/ (accessed September 14,
2010).

31. Deepwater Horizon Incident Joint Information Center, "Transcript—
Press Briefing by National Incident Commander Admiral Thad Allen,"
Deepwater Horizon Unified Command website, August 27, 2010, www
.deepwaterhorizonresponse.com/go/doc/2931/886987/ (accessed September 14,
2010).

## Chapter 10. Going for the Kill: Firefighting and Relief Wells

1. Boots & Coots International Well Control, Inc., "History," Boots & Coots
website, www.bootsandcoots.com/members/history.html (accessed September 14,
2010).

2. Adair Enterprises, Inc., "Red Adair Biography: His Story," Adair Enterprises,
Inc., website, www.redadair.com/hisstory.html (accessed September 14, 2010).

3. Incident News, "Ekofisk Bravo Oilfield," *Incident News*, Emergency Response
Division, Office of Response and Restoration, National Oceanic and
Atmospheric Administration, www.incidentnews.gov/incident/6237 (accessed
September 14, 2010).

4. Incident News, "Ixtoc 1," *Incident News*, Emergency Response Division,
Office of Response and Restoration, National Oceanic and Atmospheric
Administration, www.incidentnews.gov/incident/6250 (accessed September 14,
2010).

5. Maurice Chittenden and Richard Ellis, "From the Archive: Texan Daredevil
Enters Piper Alpha Inferno," *Sunday Times*, July 10, 1988, www.timesonline
.co.uk/tol/news/uk/article7061003.ece (accessed September 14, 2010).

6. Captain Staff, "Deepwater Horizon—Are Emergency Support Vessels Needed?"
Captain website, July 27, 2010, gcaptain.com/maritime/blog/deepwater-horizon
-are-emergency-support-vessels-needed?16404&utm_source=feedburner&utm_
medium=feed&utm_campaign=Feed%3A+Gcaptain+%28gCaptain.com%29
(accessed September 14, 2010).

7. John Wright Company, "John W. Wright," biography from the John Wright
Company website, www.jwco.com/companyinfoframe.htm (accessed September
14, 2010).

8. Trade and Environment Database, "The Economic and Environmental Impact
of the Gulf War on Kuwait and the Persian Gulf," Trade and Environment
Database website, December 1, 2000, www1.american.edu/ted/kuwait.htm
(accessed September 14, 2010).

9. William Maclean, "Volcano Chaos: A Pointer to Potential Iran/Gulf Smoke
Disruption?" Reuters, April 28, 2010, blogs.reuters.com/global/2010/04/28/

volcano-chaos-a-pointer-to-potential-irangulf-smoke-disruption (accessed September 14, 2010).

10. Boots & Coots International Well Control, Inc., "History," Boots & Coots website, www.bootsandcoots.com/members/history.html (accessed September 14, 2010).

11. John Wright Company, "Case Histories: North Sea 1988," John Wright Company website, www.jwco.com/casehistoryframe.htm (accessed September 14, 2010).

12. Adair Enterprises, Inc., "Red Adair Biography: His Story," Adair Enterprises, Inc., website, www.redadair.com/hisstory.html (accessed September 14, 2010).

13. Monica Hatcher, "Disaster in the Gulf: Master Driller Aims to Bring Relief to Oil Spill," *Houston Chronicle*, July 11, 2010 www.chron.com/disp/story.mpl/business/7103571.html (accessed September 14, 2010).

14. Jacqui Goddard, "BP Oil Spill: The Red Adair of Relief Well Drilling Says 'No Doubt About Successful Outcome,'" *Daily Telegraph*, July 11, 2010, www.telegraph.co.uk/finance/newsbysector/energy/oilandgas/7883286/BP-oil-spill-the-Red-Adair-of-relief-well-drilling-says-no-doubt-about-successful-outcome.html (accessed September 14, 2010).

15. Sterling Gleason, "Slanted Oil Wells," *Popular Science* 124, no. 35 (May 1934).

16. Julia Cauble Smith, "Conroe Oilfield," *Handbook of Texas Online*, www.tshaonline.org/handbook/online/articles/CC/doc2.html (accessed September 14, 2010).

17. Kristin L. Wells, "Technology Solves 1933 Oilfield Crisis—Conroe Crater," *Well Servicing Magazine*, wellservicingmagazine.com/technology-solves-1933-oilfield-crisis---conroe-crater (accessed September 14, 2010).

18. Jacqui Goddard, "BP Oil Spill: The Red Adair of Relief Well Drilling Says 'No Doubt About Successful Outcome,'" *Daily Telegraph*, July 11, 2010, www.telegraph.co.uk/finance/newsbysector/energy/oilandgas/7883286/BP-oil-spill-the-Red-Adair-of-relief-well-drilling-says-no-doubt-about-successful-outcome.html (accessed September 14, 2010).

## Chapter 11. The Politics of Offshore Drilling

1. Kate Sheppard, "The Curse of Bush," *Washington Post* Ezra Klein blog, May 28, 2010, voices.washingtonpost.com/ezra-klein/2010/05/the_curse_of_bush.html (accessed September 14, 2010).

2. George W. Bush, "Executive Order 13212—Actions to Expedite Energy-Related Projects," Executive Office of the President, National Environmental Policy Act Archive, May 18, 2001, ceq.hss.doe.gov/nepa/regs/eos/eo13212.html (accessed September 14, 2010).

3. Bill Moyers, "Science and Health: Wilderness at Risk," *PBS Now with Bill Moyers*, January 4, 2004, www.pbs.org/now/science/rockymtnfront.html (accessed September 14, 2010).

4. Juliet Eilperin, "US Exempted BP's Gulf of Mexico Drilling from Environmental Impact Study," *Washington Post*, May 5, 2010, www.washingtonpost.com/wp-dyn/content/article/2010/05/04/AR2010050404118.html (accessed September 14, 2010).

5. Ian Urbina, "US Said to Allow Drilling Without Needed Permits," *New York Times*, May 13, 2010, www.nytimes.com/2010/05/14/us/14agency.html? r=1&hp (accessed September 14, 2010).

6. Ibid.

7. Abraham Lustgarden, "Whistleblower Sues to Stop Another BP Rig from Operating," *ProPublica*, May 17, 2010, www.propublica.org/article/whistleblower -sues-to-stop-atlantis-bp-rig-from-operating (accessed September 14, 2010).

8. David Whitten, "Oil, Gas Royalty-in-Kind Program to End, Salazar Says (Update 3)," *Bloomberg News*, September 16, 2009, www.bloomberg.com/apps/ news?pid=newsarchive&sid=amtN_wUEgr2o (accessed September 14, 2010).

9. C. Stephen Allred, "Oil and Gas Royalties: MMS's Oversight of Its Royalty-in-Kind Program Can Be Improved Through Additional Use of Production Verification Data and Enhanced Reporting of Financial Benefits and Costs," US Government Accountability Office, letter to congressional requesters, September 26, 2008, www.gao.gov/new.items/d08942r.pdf (accessed September 14, 2010).

10. David Whitten, "Oil, Gas Royalty-in-Kind Program to End, Salazar Says (Update 3)," *Bloomberg News*, September 16, 2009, www.bloomberg.com/apps/ news?pid=newsarchive&sid=amtN_wUEgr2o (accessed September 14, 2010).

11. US Department of Justice, "Former Interior Deputy Secretary Steven Griles Sentenced to 10 Months in Prison for Obstructing US Senate Investigation into Abramoff Corruption Scandal," US Department of Justice website, June 26, 2007, www.justice.gov/opa/pr/2007/June/07_crm_455.html.

12. Al Kamen, "Jack Abramoff's New Gig Is Pizza Man," *Washington Post* Voices blog, June 22, 2010, voices.washingtonpost.com/44/2010/06/report-abramoffs -new-gig-is-pi.html (accessed September 14, 2010).

13. Jim Tankersley, "Federal Report Slams Drilling Instructors," *Los Angeles Times*, May 25, 2010, http://articles.latimes.com/2010/may/25/nation/la-na-oil -company-gifts-20100526 (accessed September 17, 2010).

14. US Department of the Interior, "Salazar Launches Safety and Environmental Protection Reforms to Toughen Oversight of Offshore Oil and Gas Operations," US Department of the Interior Press Release, May 11, 2010, www.doi.gov/news/ pressreleases/Salazar-Launches-Safety-and-Environmental-Protection-Reforms -to-Toughen-Oversight-of-Offshore-Oil-and-Gas-Operations.cfm (accessed September 14, 2010).

15. Associated Press, "Drilling Moratorium Depends on Industry, Michael Bromwich Says," *New Orleans Times-Picayune*, September 10, 2010, http://www.nola.com/ news/gulf-oil-spill/index.ssf/2010/09/drilling_moratorium_depends_on.html (accessed September 17, 2010).

16. Harry Weber, "End of Ban on Deep-Water Drilling Depends on Industry Compliance, US Says," *Fort Worth Star-Telegram* via the Associated Press, September 10, 2010, www.star-telegram.com/2010/09/10/2459055/end-of -ban-on-deep-water-drilling.html (accessed September 14, 2010).

17. Speaker Nancy Pelosi, "Current Legislation: The CLEAR Act," website of Nancy Pelosi, Speaker of the US House of Representatives, www.speaker.gov/ newsroom/legislation?id=0399 (accessed September 14, 2010).

18. Matthew Iglesias, "It's Bush's Oil Spill," *Daily Beast*, May 14, 2010, www.thedailybeast.com/blogs-and-stories/2010-05-14/its-bushs-oil -spill/?cid=bs:featured1 (accessed September 14, 2010).

19. Juliet Eilperin, "US Exempted BP's Gulf of Mexico Drilling from Environmental Impact Study," *Washington Post*, May 5, 2010, www.

washingtonpost.com/wp-dyn/content/article/2010/05/04/AR2010050404118.
html (accessed September 14, 2010).

## Chapter 12. The Aftermath: What Do We Do Now?

1. Larry H. Flak, "Ultra-Deepwater Blowouts—How Could One Happen," *Offshore Magazine*, January 1, 1997, www.offshore-mag.com/index/article-display/23675/articles/offshore/volume-57/issue-1/departments/drilling-production/well-control-ultra-deepwater-blowouts-how-could-one-happen.html (accessed September 14, 2010).

2. Joe Leimkuhler, *Drilling for Oil: A Visual Presentation of How We Drill for Oil and the Precautions Taken Along the Way*, presented at the Aspen Ideas Festival, Aspen, Colorado, July 2010.

3. BP Incident Investigation Team, *Deepwater Horizon Accident Investigation Report*, BP Accident Investigation (Houston: BP, 2010), page 193.

4. Kent Wells, transcript of technical briefing conference call moderated by Marcella Christophe, July 13, 2010, BP website, www.bp.com/liveassets/bp_internet/globalbp/globalbp_uk_english/incident_response/STAGING/local_assets/downloads_pdfs/kent_wells_technical_briefing_transcript_13_07_0730CDT.pdf (accessed September 15, 2010).

5. Deepwater Horizon Joint Investigation, "Statement from National Incident Commander Admiral Thad Allen on Well Integrity Test," Deepwater Horizon Unified Command website, July 12, 2010, http://www.deepwaterhorizonresponse.com/go/doc/2931/776891/ (accessed September 17, 2010).

6. Kent Wells, conference call moderated by Daren Beaudo, July 19, 2010, BP website, http://www.bp.com/liveassets/bp_internet/globalbp/globalbp_uk_english/incident_response/STAGING/local_assets/downloads_pdfs/BP_technical_audio_07192010.pdf (accessed September 14, 2010).

7. Dr. Susan Shaw, *The Oil Spill's Toxic Trade-Off*, presented at the TEDxOilSpill conference, Washington, DC, June 28, 2010, www.ted.com/talks/susan_shaw_the_oil_spill_s_toxic_trade_off.html (accessed September 14, 2010).

8. Noelle Straub, "Interior Unveils Plan to Split MMS into 3 Agencies," *New York Times*, May 20, 2010, www.nytimes.com/gwire/2010/05/20/20greenwire-interior-unveils-plan-to-split-mms-into-3-agen-72654.html (accessed September 14, 2010).

9. American Petroleum Institute, "Moratorium Causes Significant Harm to Economy, API Looks to Appeals Court," American Petroleum Institute Newsroom website, July 8, 2010, www.api.org/Newsroom/moratorium-harm-econ.cfm (accessed September 14, 2010).

10. Jim Berard, "Foreign Vessel Operations in the US Exclusive Economic Zone," US House of Representatives Committee for Transportation and Infrastructure press release, June 17, 2010, transportation.house.gov/News/PRArticle.aspx?NewsID=1253 (accessed September 14, 2010).

11. Deepwater Horizon Joint Investigation, "Deepwater Horizon Joint Investigation Hearing Transcript—May 11, 2010," Deepwater Horizon Unified Command website, June 9, 2010, www.deepwaterinvestigation.com/external/content/document/3043/621931/1/Deepwater%20Horizon%20Joint%20Investigation%20Transcript%20-%20May%2012 (accessed September 14, 2010).

12. Jad Mouawad, "4 Oil Firms Commit $1 Billion for Gulf Rapid-Response Plan," *New York Times*, July 21, 2010, www.nytimes.com/2010/07/22/business/energy -environment/22response.html (accessed September 14, 2010).

13. American Petroleum Institute, "Subsea Well Control and Oil Spill Response Industry Task Forces Briefing Paper 9.07.10," Scribd, September 7, 2010, www .scribd.com/doc/37123432/Subsea-Well-Control-and-Oil-Spill-Response -Industry-Task-Forces-Briefing-Paper (accessed September 14, 2010).

14. Joe Flaherty and Mary Pat Stevens, "Oil Industry Cleanup Organization Swamped by BP Spill," *Washington Post*, June 29, 2010, www.washingtonpost .com/wp-dyn/content/article/2010/06/29/AR2010062905384.html (accessed September 14, 2010).

# | ABOUT THE AUTHOR |

Bob Cavnar, an oil and gas industry executive, is the founder of the Daily Hurricane, a website covering news and views about politics, energy, health, science, education, current affairs, and national and international news. A thirty-
year veteran of the oil and gas industry, Cavnar has deep experience in drilling and production operations, start-ups, turnarounds, and manage- ment of both public and private companies. He is currently chief execu- tive officer of Luca Technologies, which harnesses natural processes to produce natural gas sustainably. Previously, he was president and chief executive officer of Milagro Exploration—a large, privately held oil and gas exploration firm based in Houston, Texas, with operations along the Texas, Louisiana, and Mississippi Gulf Coast, and offshore in the Gulf of Mexico. Prior to that, he served as chairman, president, and chief executive officer of Mission Resources, where he led the recovery of the company's financial strength and prominence within the independent oil and gas industry with operations in South and East Texas, South Louisiana, offshore Gulf of Mexico, offshore California, and Southeast New Mexico.

Cavnar blogs regularly at the Huffington Post and the Daily Hurricane and is widely acknowledged as a voice for oil and gas indus- try reform. Throughout the BP oil spill in the Gulf of Mexico, he appeared regularly on MSNBC's *Countdown with Keith Olbermann*, analyzing the unfolding disaster and explaining the technical aspects of the response. He also appeared frequently on other news programs, including *The Rachel Maddow Show*, *Hardball with Chris Matthews*, *The Ed Schultz Show*, *MSNBC Live*, *The Today Show*, and several other news shows from NBC, BBC, CBC, BNN, and Al Jazeera. Cavnar holds a Master of Business Administration degree from Southern Methodist University and completed the Program for Management Development at the Harvard Business School. He lives in Houston, Texas.

# | INDEX |